KRISTEN ACRI, OTR/L

Félicie D. Affolter

Perception, Interaction and Language

Interaction of Daily Living:
The Root of Development

Foreword by Ida J. Stockman

With 544 Figures

Springer-Verlag Berlin Heidelberg New York
London Paris Tokyo Hong Kong Barcelona

Félicie D. Affolter, Ph. D.
Via Clavaniev

CH-7180 Disentis

Translation from the German Edition
Wahrnehmung, Wirklichkeit und Sprache
(Wissenschaftliche Beiträge aus Forschung, Lehre und Praxis zur Rehabilitation
behinderter Kinder und Jugendlicher; 4)

© 1987 Neckar-Verlag, Villingen-Schwenningen

ISBN 3-540-51150-4 Springer-Verlag Berlin Heidelberg New York
ISBN 0-387-51150-4 Springer-Verlag New York Berlin Heidelberg

Library of Congress Cataloging-in-Publication Data
Affolter, Félicie D., 1926- [Wahrnehmung, Wirklichkeit, und Sprache. English] Perception, interaction and language : interaction of daily living : the root of development / Félicie D. Affolter ; foreword by Ida J Stockman. p. cm. Translation of: Wahrnehmung, Wirklichkeit und Sprache. Includes bibliographical references.
ISBN 0-387-51150-4 (U. S.: alk. paper)
1. Perceptual-motor learning. 2. Problem solving in children. 3. Language disorders in children. 4. Movement therapy for children. I. Title. BF295.A3213 1991 153.1'524-dc20 90-9718

This work is subject to copyright. All rights are reserved, whether the whole or part of the material is concerned, specifically the rights of translation, reprinting, reuse of illustrations, recitation, broadcasting, reproduction on microfilms or in other ways, and storage in data banks. Duplication of this publication or parts thereof is only permitted under the provisions of the German Copyright Law of September 9, 1965, in its current version, and a copyright fee must always be paid. Violations fall under the prosecution act of the German Copyright Law.

© Springer-Verlag Berlin Heidelberg 1991
Printed in Germany

The use of general descriptive names, registered names, trademarks, etc. in the publication does not imply, even in the absence of a specific statement, that such names are exempt from the relevant protective laws and regulations and therefore free for general use.

Product Liability: The publisher can give no guarantee for information about drug dosage and application thereof contained in this book. In every individual case the respective user must check its accuracy by consulting other pharmaceutical literature.

Typesetting: Appl, Wemding
2121/3145-543210 - Printed on acid-free paper

Foreword

Perception, Interaction and Language is essentially about the nature of adaptive learning in both its normal and abnormal forms. A major premise is that verbal and nonverbal learning depend critically on *interaction* with the environment via daily life problem solving activity. This new perspective is not entirely unfamiliar. But, a fundamentally new perspective is offered by the additional premise that tactile-kinesthetic perception associated with problem solving experiences provides the primary basis for interaction and hence, is the root of adaptive learning and development.

Although decades have passed since Jean Piaget first proposed a model of cognitive learning and development that builds on sensorimotor processes, little seems to be known about how human actions are actually translated into adaptive behavior. We are keenly reminded of this lack of knowledge when dealing with persons who present abnormal performance. The failure to learn language, in particular, can create a devastating handicap. Therefore, it is not surprising that systematic attention has been given to research and clinical programs for the language disordered. But, what have we learned so far about abnormal language? This book commands our attention at a time when we know more about how to describe abnormal language characteristics than how to explain or remedy them. We continue to be puzzled by the severely impaired whose behavioral profiles neither fit typical diagnostic categories nor respond to conventional therapy practices. What we know, however, is that language can develop abnormally or be absent altogether, despite an outwardly normal physical appearance and the ability to see, hear, taste, smell and even move the body. We know that the same person can have difficulty not only learning to speak, read and write, but also in coping with such nonverbal daily life activities as cooking, going to the store, putting on a shoe, and so on.

This book also appears at a time when we have become disenchanted with the promises of strict behaviorist approaches to language intervention. Expanding knowledge about the nature of language and its relationship to other behavioral domains has made us aware that intervention must encompass more than the traditional focus on vocabulary meaning, articulation and grammatical forms. We now sense that therapy must be modeled in ways that do not isolate the teaching of language from actual world experience in both its verbal and nonverbal contexts. But, existing eclectic approaches to intervention suggest that we are not yet clear about how to model intervention so that it accommodates the dynamic nature of behavior in actual world contexts.

Nonetheless, our approaches reflect an emerging sense that it is not enough just to hear language in order to learn it. Given current belief in the importance of relating language forms to action events, a focus on tactile-kinesthetic processes should be inescapable. But, this has not been the case. Moreover, the clinician gives attention to action events in the virtual absence of information about the nature of tactile-kinesthetic perception or its relation to problem solving activity and the acquisition of complex verbal and nonverbal behavior. It is even more unfortunate that leading scholars are not yet asking questions about the possible importance of this sensory-perceptual system for the development of higher order behavioral function.

Perception, Interaction and Language makes clear that Dr. Affolter and her co-workers have done more than ask the question. The unmistakable foregrounding of tactile-kinesthetic sensory input in a theory of learning represents bold new reflections about the nature of perceptual-cognitive mechanisms that support the learning of complex skills. This perspective did not develop overnight. It is the outcome of almost two decades of clinical observation and synthesis of research findings in the areas of psychophysics, psycholinguistics, developmental and cognitive psychology. What began as a response to the practical problem of treating children with abnormal verbal and nonverbal behavior has evolved into a theoretical perspective on learning which is broadly relevant to understanding the role of perception in both normal and abnormal behavior by children and adults.

A natural progression of ideas unfolds in the book's three major parts. Part I introduces a scaffold for the model of learning by taking us on a journey through the earliest stages of normal perceptual-cognitive development. The numerous examples of children's behavior provide a flavor of the painstakingly detailed observations that stimulated hypotheses about perceptual development which are still being tested by Dr. Affolter and her co-workers in their research and clinical work.

The close attention to observational detail is continued in the numerous illustrations of Part II which validate the model of learning proposed in Part I by painting a picture of abnormal behavior. These descriptions of clinical children's naturalistic behavior in routine daily life activity contrast sharply with the more familiar ones that result from structured elicitation tasks.

Part III has refreshing appeal to those who are tired of theoretical arguments that do not translate directly into a concrete intervention plan. The use of guided movement in a problem solving context to facilitate the acquisition of both verbal and nonverbal skill departs radically from traditional approaches to therapy on many fronts. Although the general notion of guided movement is not mysterious to any of us who have made a natural helping response to someone who has difficulty performing a task, the book reveals that guided movement therapy is hardly a simplistic undertaking. The insightful observations of Dr. Affolter and her co-workers elevate guided movement to the level of a highly specialized skill requiring very specific knowledge about how to move in the service of learning.

The ideas presented in this book have the potential of reorienting our thinking about tough clinical problems in a fresh and compelling way. *Perception, Interaction and Language* should challenge the reader to ask new questions about the role of perception in normal and abnormal learning that extend beyond the well studied auditory and visual sensory modalities. It should encourage inquiry about the tactile-kinesthetic system, a relatively unexplored frontier that should be understood in our quest to explain the nature of human learning.

Ida J. Stockman
Michigan State University

Contents

Part I: Living in a Wirklichkeit 1

A.   The Wirklichkeit as It Is 5

1   The Surroundings Begin to Take Shape 5
1.1 The World Is Here: I Touch the World
     and the World Touches Me 5
1.1.1 The Stable Support 6
1.1.2 The Stable Side 8
1.2 The World Becomes a Surrounding World:
     The World Embraces Me; I Embrace the World 8
1.2.1 The Niche – The World Embraces Me 8
1.2.2 The Object: I Embrace the World 16
1.3 I Perceive the World Around Me: I Embrace It 22
1.3.1 I See or Hear or Feel 22
1.3.2 From Feeling to Feeling and Seeing 24
1.3.3 From Looking to Taking 25
1.4 The Surroundings Become Familiar 28
1.4.1 Unfamiliar – I Jerk Away 28
1.4.2 Searching for New Events 29
1.4.3 One Hand – Then Two – And Still a Unity 31
1.4.4 The Multitudinous Ways of Touching and Releasing 33

2   The Wirklichkeit: Perceiving and Acting Upon 38
2.1 Cause and Effect 38
2.1.1 I Set Things in Motion 38
2.1.2 I Separate and I Bring Together 41
2.2 I Explore Neighboring Relationships Through Feeling 46
2.2.1 I Take out and Put in 46
2.2.2 It Goes Through and Then Where Is It? 50
2.2.3 It Disappears and I Find It Again 52
2.3 The Wirklichkeit Becomes Familiar to Me as It Is 55
2.3.1 I Grasp a Multitude of Causes... 55
2.3.2 ...and Effects – Outdoors and Indoors 60

| B. | The Wirklichkeit Can Be Changed | 69 |

1	The Wirklichkeit as I Want It to Be	69
1.1	I Restore the Wirklichkeit	69
1.2	I Behave in an Orderly Way	69

2	Events of Daily Living Change the Wirklichkeit	70
2.1	I Help in Daily Living Events	70
2.2	I Continue to Perceive and Act Upon	72
2.3	I Can Do It Myself	77

Part II: Failing in a Wirklichkeit 83

1	Those Around Them Notice: They Are Deviant	87
1.1	They Are Either Too Hectic or Too Quiet	88
1.2	They Talk Incessantly	89
1.3	They Are Called Aggressive	92
1.4	They Are Labeled Ill-Mannered	94

2	We Observe: They Have It and Yet They Don't Have It	96
2.1	They Know About the Rules of Touching, but Where Is the World Around Them?	96
2.1.1	They Withdraw from Touching, Become Tense and Look Away	96
2.1.2	They Know About the Rules of the Stable Support and the Side	104
2.1.3	They Have Two Hands but Often Use Only One...	115
2.1.4	...and Don't Succeed in Embracing Things	118
2.2	They Know About the Rules of Acting Upon – But Where Is Their Changing of the Surroundings?	122
2.2.1	They Take Off – But How?	122
2.2.2	And Where Is the Neighborhood?	128
2.2.3	The Sequence – When Something Is Missing or When You Cannot Go Back	131
2.2.4	When Only the Moment Exists...	133
2.2.5	...and Causative Actions Do Not Correspond to the Situations...	135
2.2.6	...Then the Wirklichkeit Slips Away	137

3	What Happens When There Is a Lack of Tactile-Kinesthetic Information?	139
3.1	They Search for Information	139
3.1.1	They See and Hear	139
3.1.2	They Receive Tactile-Kinesthetic Information	141

3.2	When Information Is Deviant...	144
3.2.1	...Then Problems Are Recognized but Not Solved...	145
3.2.2	...and the Surroundings Are Still Unfamiliar	145
3.2.3	They Hardly Know What Is Happening Around Them	146
3.2.4	The World Does Not Become a Surrounding World	147
3.3	The Limitation of Capacity	151
3.3.1	What Are the Consequences of a Limited Capacity?	151
3.3.2	When I Can Order Information	152
3.3.3	What Happens When I Am in Search of Tactile-Kinesthetic Information...	153
3.3.4	...and the Competence Does Not Become Performance?	154

Part III: Learning in a Wirklichkeit . . . 157

A.	**Problem Solving Events Are the Root of Development**	160
1	Development Occurs with a Surprising Regularity	161
2	But What Happens When Children Fail in Perception?	161
2.1	The Development of Perceptual Performances Is Deviant	161
2.2	Developmental Performances Appear in a Different Sequence	161
2.3	Problem Solving Activities Are Deviant	162
3	How Can We Represent Development?	162
3.1	Interpreting the Results	162
3.2	The Model of Development	163
4	We Cannot Simply Wait	164
4.1	We Should Not Practice Skills...	165
4.2	...but Should Begin with "Problem Solving Events" and Mediate the Corresponding Tactile-Kinesthetic Information	165
B.	**Problem Solving Events Can Be Felt**	166
1	I Feel and Can Change My Behavior	166
1.1	I Learn from Tactile-Kinesthetic Experience	166
1.1.1	What I Feel Is Unfamiliar to Me	170
1.1.2	I Feel and It Becomes Familiar; Now I Can Also Look at It	170
1.1.3	I Feel and Look – I Look and Feel	174
1.1.4	I Recognize What I Feel and Continue the Movements; I Anticipate Them	175
1.2	I Touch and Allow for Touching	178
1.2.1	I Guide from Behind	178
1.2.2	They Have Two Hands, a Mouth, and a Body	180

2	I Feel and Act upon	189
2.1	I Feel the Wirklichkeit	189
2.1.1	The Wirklichkeit Includes the Surroundings,...	189
2.1.2	...and Changes in Resistance Are Needed	192
2.2	They Know About the Rules...	195
2.2.1	...of Touching...	196
2.2.2	...and Acting Upon	197
2.3	And Now I Can Change the Surroundings	203
2.3.1	When to Use Hands and When to Use Tools	210
2.3.2	Everything Happens in Its Own Time	220
2.3.3	Am I Allowed to Break That?	220
3	I Understand Problem Solving Events of Daily Living	221
3.1	Learning Begins with Understanding	221
3.1.1	Understanding – What Is Meant?	221
3.1.2	I Work with Children and Adults on Their Level of Understanding	222
3.2	Problems Arise All the Time	227
3.2.1	Problem Solving Is Thrilling	227
3.2.2	Difficulties Can Be Overcome; How Good It Is to Have Difficulties	233
3.2.3	The Solving of the Problem Is Important, Not the Product	234
3.2.4	I Have Solved the Problem by Myself	235
3.2.5	When Do I Talk?	236
3.3	Guided Through Daily Living	237
3.3.1	I Guide When Problems Arise in Daily Living...	237
3.3.2	...or I Have to Plan Problem Solving Events	241
3.3.3	Feeling Is Very Difficult	242
3.3.4	A Break in Guiding Allows Time for Thinking	243
C.	**Tactile-Kinesthetic Experiences with Solving Problems of Daily Living Are Interiorized**	244
1	Production Begins	245
1.1	From Anticipation to Producing	245
1.2	Initial Acts of Production Are Eagerly Awaited – But?	246
1.3	It Is Good to Have Habits	248
1.4	Habits Do Not Make Progress Happen	248
1.4.1	Habits Are Rigid	248
1.4.2	Habits Break Down When Situations Change	249
1.5	The Gift of Curiosity – How to Break Through Habit Formation	250
1.5.1	If Information Is Restricted, How Can It Be Expanded?	251
1.5.2	Feeling Causes and Effects	255
1.6	The Way Back to the Main Road Is Found by Using Detours	256
2	Return to the Problem Solving Events of Daily Living – Then Comes Representation	267

2.1	The Symbol Represents Problem Solving Events of Daily Living	268
2.1.1	Problem Solving Events Are Represented by Forms Which Are Ready-Made	268
2.1.2	The Forms for Representation Are Constructed	271
2.1.3	The Symbol Serves to Explore New Situations	273
2.1.4	The Picture Is a Symbol	275
2.2	Problem Solving Events in Daily Living and Then – Language Performance	276
2.2.1	Children Relate Verbal Forms to Problem Solving Events They Experience	276
2.2.2	Problem Solving Events in Daily Living Are the Basis for Deep Structure and for Surface Structure	279
2.3	When the Root Is Sick...	281
2.3.1	...the Symbolic Behavior Is Deviant...	281
2.3.3	...the Deep Structure and Surface Structure Are Deviant	283
2.4	What Can Be Done?	284
2.4.1	Something Can Always Be Done	284
2.4.2	First Comes Tactile-Kinesthetic Input from Problem Solving Events of Daily Living and Then Comes Representation	285

D. Conclusions . . . 292

1	Problem Solving Events Can Be Considered the Root of Development	292
1.1	Possibilities of Application Have a Variety of Forms	292
1.2	The Application of the Therapeutic Model Involves a Whole Circle of Persons	293
2	Our Knowledge About the Root of Development Is Still Limited	294
2.1	Extension of Longitudinal Research Is Needed	294
2.2	Extension of Cross-Sectional Research Is Needed	295

Closing Remarks . . . 296

Glossary: Wirklichkeit . . . 299

References . . . 301

Subject Index . . . 303

Introduction

For many years I worked on this book. It began with *normal* children. At the College of Education, we were told what subjects we should teach normal children as they go to first, second, third,...eighth grades. As we discussed the "What," I began to ask, "Why"? In order to answer that question, I started investigating normal children's development. Jean Piaget became my teacher (Affolter, 1981), and I learned much from him, especially the importance of observing children in experimental as well as in natural situations.

As a young teacher, I came in contact with *profoundly deaf* children who did not acquire oral language spontaneously. I began to reflect on how language could be taught and was amazed at how normal these deaf children were. In my experience I learned that thinking can occur independently of language (Piaget, 1963; Furth, 1966), and that children without hearing and with very little language and speaking ability can develop normally, both cognitively and emotionally. How could this be explained? This question became the focus of my further graduate study and clinical observation.

With increasing frequency, I encountered other children, who, like the deaf children, also failed in *language development.* Because these children could hear, the reasons for their failure could not be attributed to a hearing loss. I soon noticed that their difficulties with language development was not their only problem. They also failed in some nonverbal tasks. More careful observations led me to assume that their difficulties began at the prelinguistic level of development (Affolter, 1972).

In comparing the latter group of language disordered children to those who were profoundly deaf, some questions arose: Why do language disordered children behave differently from deaf children? Why can't they learn language adequately in spite of their good hearing? No one could answer these questions; there were no answers in the literature either.

Adult patients followed. They were people who had developed normally prior to suffering a brain lesion. After brain damage, whatever its cause, behavior problems began for them. These patients were sometimes young, sometimes older, and sometimes even geriatric. They failed to solve basic problems in daily life and they failed to communicate adequately with others. One could observe a regression in performance that had been learned during developmental stages. How could these people be helped? Available knowledge about regression in performance was as limited as that about development.

Therefore, a team of co-workers and I undertook *research projects* which were sponsored by the Swiss National Science Foundation. Investigations were based on clinical work with profoundly deaf and language disordered children and with brain damaged adults. Different clinical programs were applied and the behavioral responses of both children and adults were observed. The observations were compaired, grouped, analyzed, and evaluated. As new questions were raised and new hypotheses were formulated, the clinical programs and situations were modified. As the questions became more focused, we began systematic cross-sectional studies while expanding the longitudinal observations.

It was, and still is, like a hike to an unfamilair mountain. One has to search for a path. When the path is found, one can continue on it for some time – to the next crest. There one can rest, look around, and orient oneself. "How far did I go? Where is the next step? How will I proceed"? We have not as yet arrived at the highest point of the mountain from where it would be possible to survey the whole region. We still need to search further for the continuation and the precise direction of the mountain path. But we believe that an important summit has been reached, high enough to enable us to reflect on what we have done so far. The path is safe, too, so others are invited to join us – to pause with us and to progress with us – as we travel further up the mountain.

This book contains *reflections* which also encompass the *knowledge we have to date*. The problems that will be described arose in practical work with normal and disordered children and adults. The problems refer to development – its disorders and regressions. All of these are interrelated. To understand disorders and regression in development requires an understanding of normal development. Increasing knowledge in one field enhances understanding in the other. This means we will continually compare the behavior of normal children with the behavior of disordered ones and the behavior of normal adults with the behavior of adults who are brain damaged.

The text includes observations chosen from a multitude of possible ones. They are especially important for an understanding of the root of development and its disorders as well as for describing the processes of learning. The examples are used to describe points of view which are rarely found in present literature. Through the examples we will emphasize the importance of observing. The ability to observe accurately becomes more imperative for us when we consider the growing number of theories concerning development and education. These theories are becoming increasingly disturbing, especially in view of the effects of the so called "liberal" ways which often result in "anti-authoritarian" behavior.

The reflections in this book have implications of social and sociopolitical importance. They are significant for home and school organizations and are useful to insurance companies in setting up policies and programs dealing with accident cases. Only some of these implications will be mentioned at the end of this book. To describe all of them would require a subsequent book.

I would like to *thank* the many people who have helped us reach our present level of knowledge.

Thanks to all the parents, educators, physicians, therapists, teachers, etc., who came to ask my clinical staff and me for help. They placed their confidence in us, encouraged us and supported our clinical work and research. Among those that I would like to especially thank are the medical doctors, W. Zinn, E. Stricker, Professor G. Weber, Professor E. Gautier, and H. Städeli.

My special thanks go also to the Swiss National Science Foundation (SNF projects nos. 3.237.69, 3.448.70, 3.902.72, 3.2050.73, 3.504.75, 3.711.76, and 3.929.078) which sponsored our research for 10 years, the Swiss Association for cerebral palsied children, and the Arbeitsgemeinschaft für die Probleme Wahrnehmungsgestörter.

I am also grateful to the Center for Research in Learning, Perception and Cognition at the University of Minnesota in the United States, whose staff has repeatedly supported and stimulated our research.

Thanks to all the persons who have helped with the English translation, especially to Judith Gloystein and William Franklin. Also, thanks to Ida Stockman, Pat Broen, Gerhard Siegel, and Herb Pick, Jr. I am most grateful for their continuous effort to understand my writings and grasp my ideas as well as to find appropriate formulations in English.

I would not have been able to write this book without the long-term support of a team of co-workers. My special thanks go to Walter and Helen Bischofberger and Doris Clausen who have sustained me professionally and morally, and who have never given up even during times of hardship and stress. Thanks to the team at the Center for Perceptual Disturbances in St. Gallen, Switzerland – those who have worked there previously and those who are working there presently. They never became tired of adding new observations, changing traditional kinds of therapies, or trying out new approaches. The same holds true for the teachers and educators of the Sonderschule für Wahrnehmungsgestörte in St. Gallen; for the physicians, therapists, and nurses in the rehabilitation Clinic Valens; and for the principal, Gisela Rolf and the instructor, Pat Davies, of the Postgraduate Study Center Hermitage attached to the Medical Center in Bad Ragaz.

Thanks to Matthias Neuweiler who painstakingly selected the pictures and arranged them with professional knowledge. Thanks also to Ruth Affolter who spent numerous hours correcting the style of the text in the German edition.

Part I
Living in a Wirklichkeit*

* see Glossary: Wirklichkeit pp. 299–300

I touch
the resistance the world offers me
through the support below me
and the sides next to me.

I change
the resistance
between my body
and the world.

I embrace
the resistance I meet on the support
with my hands and my mouth
and can thus perceive what is around me.

I cause effects
and receive effects.
In this way my surroundings
become Wirklichkeit to me

Will I learn
to change my surroundings appropriately?

From the beginning to the end of life we are placed in a *Wirklichkeit*.

We can move, but our moving is restricted by our surroundings. We seek extended space - on a lake or in the mountains. We long for freshness, for wind. At other times we desire narrowness, restriction, and warmth. We need sociability, pulsating life, and the crowd; yet, we cannot scorn being quiet or being alone for any length of time. We search for freedom, but still we need resistance.

Why are such restrictions and contrasts necessary? Are they really needed? Are they so important for the "development" of our perception? For the discovery of the Wirklichkeit? And beyond that, for the discovery of language?

I am on the mountain. A beautiful vista is before me. Range upon range of mountains extends in either direction; valleys with tiny houses are in between; cars appear as toys as they move on roads that look like small lines; people appear as tiny as ants.

I close my eyes and try to forget this distant world. Noises still reach me from below: the rattling of the motorcycles and their roaring around the curves, the impatient blowing of horns from cars, the barking of a dog, and the shrieking of a chain saw.

Lying on the rocky ground, I close my eyes again and try to forget the noises. I can feel the hardness of the stone where I am stretched out, and the warmth of the sun retained in the stone penetrates my body. I have a feeling of comfort.

Then something touches me, tickles me. I look - a butterfly sits on my bare arm. I turn over, and now I am lying on the grass. What an aromatic smell - the alpine clover!

This is the world that is *close* to me - *near* to me! I can touch it. I can move on it. I can *feel* it. I can grasp it. I can take it - it is there. It offers *resistance*.

This world that is near to me contrasts with the one that is *far away*. The world that is far away consists of colored spots I can see and sounds I can hear but *cannot touch*. It does not offer resistance when I move; therefore I cannot grasp it. I can look and listen. A short while ago I was part of all that but I wonder if the distant world is just an illusion. "Do those things - houses, people moving around, cars, the motorcycle, the chain saw - really exist"?

I close my eyes again and try to feel the world that is near and to imagine the world that is far away. I think about it. "How does it happen that I know something about my world that is close? What is it like in the world that is far away? How am I able to imagine it"?

We enter this world and experience it all around us. It surrounds us. Such an experience requires perception. In order to perceive, we use several sensory modalities. The most important, and at the same time the most complex modality, is the tactile-kinesthetic sensory system. We refer to it as "feeling". Feeling provides us with the basic information for getting acquainted with our environment. In addition to feeling we also hear, see, smell, and taste.

Whenever we speak about *perception, in its broadest sense*, we refer to the input of stimuli over the different sensory modalities which allows us to relate to the exterior world or our surroundings. However, the surroundings are not the Wirklichkeit. To get to know about the Wirklichkeit we have to know about causes and effects involved in interacting with the surroundings. How do the surroundings act upon us, and how do we react to the surroundings? How do the different parts of our surroundings act upon each other as things or persons?

To make this possible, we must be able to *perceive* in a more *narrow* sense. We can only act on something if we can *touch* it and take it. By taking something in the surroundings and acting with it, we focus our attention on it. In this way we can experience that things and persons in the surroundings exist. We perceive them.

We may be able to take hold of something we touch. Taking hold involves touching – *feeling*. Feeling is closely related to acting upon something. Whenever we think about touching something, taking hold of it, or acting upon the world – our surroundings – we must consider the feeling of it or the tactile-kinesthetic input.

How all of this is connected and develops in the child is what we will reflect on in Part I, "Living in a Wirklichkeit".

A. The Wirklichkeit as It Is

1 The Surroundings Begin to Take Shape

1.1 The World Is Here: I Touch the World and the World Touches Me

I explore the world by touching it. I move my body; I move my hands, arms, legs, and trunk. I move until I find some resistance to block my movements. This provides me with contact. The *-tact* of contact refers to a tactile input, to touching; the *con-* means with – touching with. I touch with you, and you touch with me. We stay in contact. When touching I experience resistance. This resistance is the basis for the following important discovery: Here is something that resists my movements – something that differs from my body, from me. This something is "the world," and here am "I". Thus, touching becomes the first step of an interaction which takes place between the world and myself.

L. is 6 days old. He is going to be bathed. He cries as his mother carries him to the bathtub. He kicks in the water and continues to cry. Suddenly his crying and movements stop. His body has touched the bottom of the tub.

What has happened? L. was carried through free space. He was held in his mother's arms, but the situation was unstable. The support given by her arms was mobile. L. cried. He touched the water, but the support was still moving and unstable. When his feet touched the bottom of the tub, his movement was stopped and he became quiet. The sudden change – from no resistance to total resistance – allowed him to make *contact* with the world.

We, too, have similar experiences. As soon as I touch something I feel resistance. The passage from no resistance to total resistance gives me important tactile-kinesthetic or touch information. Of significance is the *change in resistance.* I experience resistance only when it changes. The baby in the tub experienced a maximum change in resistance, i.e., from no (or minimal) change when being moved by his mother through free space and the water, to a marked change when he touched the bottom of the tub. We all need such changes in resistance to make sure of the world's and our own existence.

We assume that during the first weeks of life, babies do not know how such changes in resistance are elicited. They cannot differentiate the part their own bodies play in these events from the part played by the supporting surfaces or by the sides. The more often babies move or are moved, the more experiences they get, the better they can learn to recognize the different causes of changing resistance. When we touch a support, our bodies encounter the resistance created by *gravity* (Howard & Templeton, 1966). Experience with the regularity of this occurrence allows children to learn the "rule of support". This is expressed, for example, when babies learn to change the resistance between their bodies and a support in such a way that they can turn from lying on their backs to lying on their stomachs, or move from lying on their stomachs to the crawling position.

Besides the rule of support, other experiences of moving and touching, with the resulting changes in resistance, permit acquisition of another rule. When babies lie on a support, they will move across it by rolling, sliding, and crawling They use their arms, hands, and legs. Such movements may suddenly be stopped by another kind of resistance, a resistance on the supporting surface, but at the *side* of their bodies. Such a change in resistance at the side of their bodies is experienced in *addition* to the change in resistance between their bodies and a support. This additional change in resistance at the side does not include the factor of gravity. Regularity in experiencing this additional kind of event allows children to discover the "rule of the side." The rule of the side includes two kinds of changes in resistance: the change between the body and its support due to the effect of gravity, and at the *same time*, the change between the movement of the body along the support and the encountering of a resistance *on* that support, at the side of the body.

Thus, experiences with two rules, the rule of the support and the rule of the side, allow children to recognize that there is a world *under* their bodies, and at the same time, a world *alongside* their bodies.

The following sections briefly discuss the changes in resistance between the body, the support, and the side and emphasize their importance for development.

1.1.1 The Stable Support

Gravity binds us to a support. We continually make sure our support is a stable one. It is terrifying when a chair we are sitting on or the floor we are standing on suddenly

D., 3 years, 5 months, is walking through the woods where he discovers a pile of logs. He immediately wants to climb up and walk along the top of the logs.

Full concentration is required on this kind of support! ▶

His little brother, M., 22 months, wants to do the same thing. He also explores of the unstable condition of that support.

However, he is not as skilled as his older brother and needs a stable side; his mother's hand provides that.

collapses. Instances like these occur during earthquakes and create great panic.

In contrast, children and adults in certain leisure and sport activities like to alter changes in resistance between their bodies and their support, but they like to do so under conditions they can predict. We all have certain expectancies for the changes in resistance between our bodies and a support in specific situations. A case in point might be riding in an elevator. If I expect it to stop at the third floor, and it doesn't, the expected change in resistance has not occurred. When this happens, I experience a strange feeling in the pit of my stomach.

Children need to have many experiences with changes in resistance between their bodies and a support in order to be able to make adequate predictions. They panic when they think a support is stable and it is not.

J., 4 years, is sitting next to a mountain creek during a picnic. She has finished eating and starts to climb over some rocks. Suddenly she begins to cry. What has happened? She doesn't appear to be hurt. Actually, she only lost her balance – her support; one of the big stones on which she was climbing wobbled.

Children of many ages like to explore the stability of different kinds of support. While walking along the street, if they spy a wall, they will try to climb up and walk along it. Such changes in resistance between body and support provide them with important experiences about the world being touched.

In all of these interactions, children not only change the resistance between their bodies and a support, but also change it between their bodies, supports, and sides.

1.1.2 The Stable Side

J., 8 months, is lying on the floor. She reaches out to grasp objects. While doing this she pushes her body backward until her feet reach a resistance created by the wall or perhaps by the legs of a table. This behavior can be seen repeatedly. She is brought into a room and placed in the middle of the floor. A few minutes later she is touching the wall or a chair or a table.

R., 10 months, lifts himself up as soon as he finds a side resistance and then tries to move along on his two legs.

As soon as children or adults are on an unstable support, they will search for a stable side. When I go over a small bridge, I am grateful for the railing. As I walk down a steep staircase, I hold the railing, too. The same is true for small children. It is interesting to observe them while they are walking down a set of unfamiliar stairs. They use the information about changes in resistance between their bodies, the support, and the sides when trying to reach the next step. They slide one foot along the step they are standing on until a change in resistance is felt at the edge. Then that foot follows along the vertical side of the step, moving down until it feels the stability of the next support – the next step. Now the same sequence of movements will be repeated with the other foot.

As children increase in age, they expand their touching experiences with supports and with sides. The more extended their experiences, the more able they will be to distinguish which aspects of changing resistance are due to their own bodies and which are due to the surroundings. Thus, they gain knowledge about their bodies in *contrast* to their supports.

1.2 The World Becomes a Surrounding World: The World Embraces Me; I Embrace the World

The most extreme situation of being embraced is when the surroundings offer a stable support as well as stable sides – this is the niche.

1.2.1 The Niche – The World Embraces Me

D., 4 years, 3 months, has a new baby sister, T. He is allowed to hold her in his arms.
He is sitting down and T. is laid on his lap. The support is stable for both of them.
He puts one arm around her, and with his other arm, he holds her body tightly on his lap. In this way, there is stability – not only on the support but also at the sides.
The experience for little T: The world is holding/embracing me.
The experience for D: I am holding/embracing my world.

What a source of security! Experience with the niche situation is the basis and a point of departure for acquiring knowledge about our surroundings. This will be followed by acquiring knowledge about the Wirklichkeit – *cognitively and emotionally.*

T., 5 months, is in her buggy. She kicks her legs and waves her arms. Her whole body and both legs move first to one side and then to the other! Where will the movements be stopped? Where can she move freely?

In the beginning, babies are almost always in situations of being held all around, of being in a *niche*. They explore these situations.

M., 16 months, takes her first bath in a big tub. She cries. Then she is put into a small tub, and at once she is quiet.

Why? In the big tub M. could feel the instability of the water that was all around her body. Where does it end? Where is her body? There are no restrictions in this big tub. There is too much freedom of movement which creates fear.

The situation in the small tub is different. As soon as the body moves, it finds resistance – resistance from below and resistance from the sides. Now M. can differentiate where her body is and where her surroundings are. The tub embraces her body from the bottom and the sides. It forms a niche.

All during their growth, children either search for or create niches in order to explore and gain better knowledge about their surroundings and, at the same time, about their own bodies.

D., 11 months, finds a basin with water. How wonderful to step into that basin – to feel the resistance – at the sides and at the bottom. As a contrast, he feels the moving water.

He feels: ▶
"The world embraces me. I feel its restriction. Now I experience where I am and where my surroundings are".

D. feels secure in the niche of the basin. Consequently he can now begin a new game with his body and the water. In this way he can discover new aspects of his surroundings.
▼

M., 17 months, discovers a bucket in the kitchen. "Is it possible to step into that bucket"?

She explores the inside with her hands. Could she really step inside? There is no resistance to feel. She reaches further into the bucket. There, finally, down deep, her movements are stopped. Carefully she steps inside as her feet feel for the resistance of a support at the bottom. She holds the sides of the bucket with her hands.

Now there is room for the rest of her body. She can even sit down inside.
There is still some space between her body and the sides of the bucket.

Both arms and hands can fit into the bucket, too. Now she can't move at all. There are restrictions all around!

In this situation, she can relax in the security of her niche.

Through such experiences, children become familiar with their surroundings and, at the same time, with their own bodies. Experience follows experience.

D., 23 months, finds a wide pan. "How big is it? Can I sit in it"?
He tries it out. "Success"! He finds resistance at the bottom and resistance at the sides. He can even hold on to the sides.

Here is D., 3 years, 6 months, outside in the snow. How great to sit on the snow!
It gives way. How far?
It is almost like being in a bed – so soft – and now the restriction from below and from the sides.
"How great! I am sitting in a niche of snow"!

▲ *E., 20 months, is out for discovery. He finds one of his daddy's shoes. "How big it is! Can I slip inside"?*

The support is stable but the sides are giving way.
"Now these laces! Can I tie them"?

"How do I bring them together"? ▶

"There. Now I can begin to walk"! ▶
This is different from walking in his own shoes; with these shoes there is unstable resistance from below, from above, and from the sides.

 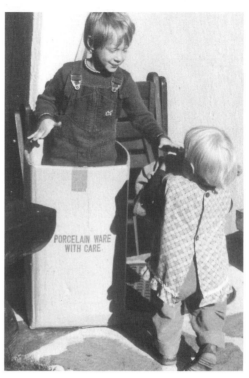

J., 3 years, 6 months, finds a big box that is open at the top.

"Can I climb inside that box? Is there also space for my teddy bear in this house"?

The situations become more complex. There is not only space for the child's own body in the niche, but also for *additional things from the surroundings*.

Children soon learn that what they see around them can also be touched and, therefore, exists. They begin to play with that knowledge. We can assume that the games of peek-a-boo and hide-and-seek are an expression of this kind of play. Children try to take away the world they see when playing. If one doesn't see things around, maybe those things cannot be touched and thus, maybe they cease to exist?
Young children begin to play peek-a-boo by covering their eyes. If they put their hands over their eyes, they no longer see what is around them. Does it still exist in spite of that? Children enthusiastically take part in this kind of play, especially if a familiar person plays it with them. Soon they discover that, instead of using just their hands to cover their faces, they can also use a blanket, and thus, close out all of their surroundings. When they remove that cover the world comes back.

The blanket over the face later becomes the cover around the body. "I don't see other people anymore and they don't see me, either".

E., 3 years, 2 months, is lying on a blanket on the grass. He begins to wrap himself up in it.
Now the blanket is all around him. He can't see anything anymore. It is dark.

▲
But maybe the things he saw before are still there?

He lifts up his head. The blanket opens, and he can look out.

He wraps the blanket around himself again.
"Now where am I?
Where is everyone else"? ▶

▼
Again E. peeks out from under the blanket. "I can see the others. Can they see me, too"?

"What if I close my eyes? I don't see them anymore. Do they see me"?

1.2.2 The Object: I Embrace the World

The movements of touching can be continued. When touching an object, our fingers move around the object until their movement is stopped. We are now holding the object. For holding, we mainly use our hands/fingers and mouths/lips. Very young babies move their arms, hands, and feet purposefully and continually when they are awake. They explore their surroundings, for example, by sliding over the support. From time to time, an arm leaves the support, is lifted up, and drops back down again. The arm may fall on an object lying there on the support, and so press the object against the support with the palm of the hand. They feel the changes in resistance caused by the support, the object, and their bodies. The search for such changes in resistance allows their fingers to move *around the object* until they cannot be moved any further. They are now *embracing* the object on the support.

T., 2 months, is lying on her back. She looks around. At the same time, her fingers move continuously over the surface of her pants, searching for changes in resistance – for something to grasp.

Holding – enclosing creates several distinguishable changes in resistance: The fingers move and embrace something until that *something* stops the movements of the fingers. To make sure they are holding an object in the hands, it is essential that they feel a *change from no or minimum resistance to total resistance*. Even young babies search for such information. While holding an object, they press their fingers around it with such force that the fingertips become white.

Fingers are not the only part of the body used for holding. Babies also use their mouths intensely for long periods of time. We refer to this as *exploration with the mouth* or mouthing.
Children put all kinds of objects into their mouths and explore the objects by making *numerous movements* with the different oral structures. For instance, they bite on an object by moving their jaws, suck it and turn it with their tongues, hold it with their lips, and lick it with their tongues.

Often children try to put the whole object into their mouths. We are surprised when we see how big babies' mouths can be. They appear to reach from one ear to the other.

T., 6 1/2 months, holds a ring with one hand. She brings it to her mouth and explores it. She touches the narrow edge with her lips and tongue.

D., 6 1/2 months, grasps whatever he can with his hands, puts it into his mouth, and explores it.

She turns the ring so that it is flat. "Can I hold it this way with my mouth"?

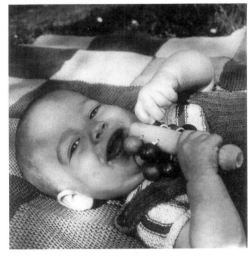

She tries to embrace it with her mouth. Using her cheeks and nose, she touches the ring. And now, she sees another object to explore – the ball!

E., 5 months, grasps objects and puts them into his mouth.
His lips embrace the objects. How far can they be put into his mouth?

 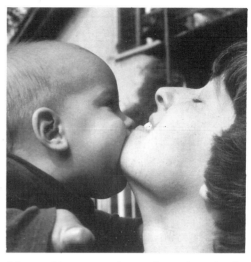

By biting he explores the resistance of these objects.

Many things can be embraced with the mouth. Such things include not only objects, but the living environment as well.

By exploring with the mouth, the coordination between lips, jaws, palate, tongue, hands, and fingers becomes very refined and highly differentiated. We can assume that such experiences are an important prerequisite for normal speaking.
Usually, children only put those things into their mouths which they have *previously touched with their hands*. A certain familiarity with the stimulus through touch appears to be required for mouthing.

▶

T., 7 months, is eating a banana. Her mother feeds her with a spoon. With her fingers she hangs on to the spoon with the banana on it. These items are quite new to her mouth. She uses the fingers of both hands. There are so many feelings connected with this activity! There are changes of resistance and changes of quality. How hard the spoon is; how soft, wet, and sticky the banana feels with her mouth, her lips, her tongue.

▲ T. accompanies the full spoon to her mouth.

"Mmm"! She puts it in her mouth.

Now it's in her mouth! ▶

Her mouth and spoon are empty.

How thrilling. The spoon is filled again!
▼

What a multitude of impressions – through the hands, the fingers, the lips, the tongue, and the cheeks!

In this way, the mouth plays an important role in the exploration of the environment.

As children develop, the *sequences of touching and holding become longer*, first with the hands and then with the mouth.

Through these kinds of experiences, the *surrounding*s – environment – are slowly shaped and become *three-dimensional.*

T., 7 months, explores a plastic strainer. ▶
She touches it...

and then tries to embrace what she has touched. ▶

She succeeds – she holds the object tightly. ▼

Afterward, because it begins to be more familiar, she embraces it with her mouth.
▼▼

To Summarize:

By touching what is around them, children pick up important information about the existence of their own bodies and limbs – as distinct from the existence of the world about them. Through innumerable interactions, they experience regularities with changes in resistance between their bodies, supports, and sides. Such regularities allow them to acquire certain rules – the *rules of touching*. These rules help children to group their experiences of touching and, in time, to make certain predictions. In this way children do not experience themselves as being in a vacuum, but rather as continually interacting with the world and experiencing the resistance of the world opposing their movements. Because of such interaction, the bodies take on conscious bodily characteristics; the bodies become their *own*, and the world slowly takes on the shape of their *surrounding world*.

1.3 I Perceive the World Around Me: I Embrace It

During the first weeks of life, babies learn to inhibit their body movements and begin to focus on a modality-specific stimulus for an increasing length of time.

1.3.1 I See or Hear or Feel

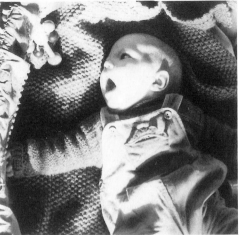

T., 2 months, looks at the toys hanging at the side of her crib. She fixes her eyes on them for a long time. ▶

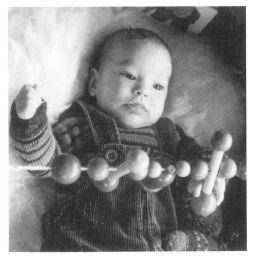

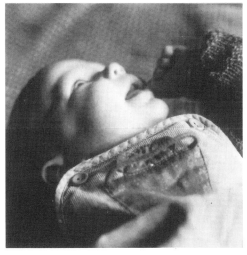

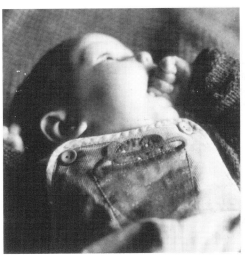

Within 2 months she has learned to follow a moving object with her eyes. They move in a smooth manner and not with the jerky movements of her first few weeks of development (Aslin, 1981).

T., 3 months, is sucking, and seems to concentrate entirely on the tactile input. Her eyes look into empty space.

Until about 3 months of age, babies are either looking or feeling or hearing. The input appears to be especially of a modality specific kind, but at 3 months, a new kind of behavior is observable. It can be most easily noticed by observing children when they are feeling and seeing.

1.3.2 From Feeling to Feeling and Seeing

*J., 2 years, 6 months, is looking at her sister, M., 3 months, who is in her crib.
"Is M. asleep"?*

*J. touches her.
M. seems to feel the touch. She opens her eyes but isn't looking in the direction of J.*

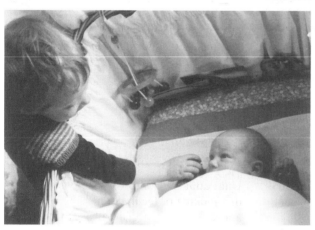

J. continues to touch M. Now M. feels the touching and looks at her.

When babies are touched they begin – with some delay at first – to also look at the source of touching.

A similar kind of behavior is elicited when babies touch something. They touch an object for a few moments, maybe even grasp it, and then begin to look at it, too.

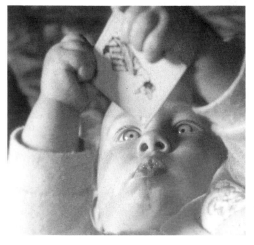

M. is 5 months old. She touches and holds a card. She looks very attentively at what her hands are doing.

The temporal delay between touching and looking becomes shorter as the experiences broaden, and the time spent looking at what the hand is doing becomes longer. This is often called hand-eye coordination (Stambak, 1963).

From this stage on, children frequently watch what their hands or bodies are doing. We can assume that in these situations children coordinate what they feel with what they see of the same event. The reference here is to the coordination of the different sensory modalities – *intermodal integration*. In the above example of eye-hand coordination we can speak about tactile-kinesthetic-visual experience.

1.3.3 From Looking to Taking

In addition to children touching their surroundings and responding by looking, other situations also become more frequent. Besides seeing something, they may attempt to touch it.

At first this performance is clumsy, but soon children's skills improve.

For the last few days, M., 7 months, has been placed on a blanket on the lawn. From this position she can feel the grass, and she tries to grasp it. With great concentration, she gazes at what she is touching.

T., 2 1/2 months, sees a ring.

 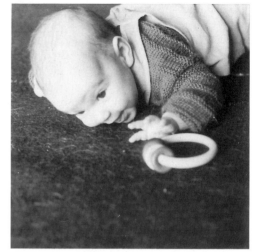

▲
*One hand is near it.
"Can I reach it"?
Her whole body stretches out.*

Her fingers touch the ring. "Can I embrace it"? It isn't possible. The ring is too far away.

T. sucks her fingers. ▶

At about 7 months, babies are quite successful at grasping. The fingers show anticipatory patterns when reaching out for an object.
▼

E., 6 1/2 months, is sitting in his chair, trying to reach out for toys.

He reaches out with his arms. In doing so, his hand opens, and his fingers spread out in anticipation of touching the toy. His eyes gaze at both the toy and his hand.

His fingers touch the object. "Can I embrace it and take it"? ▶

Here is the whole sequence of what happens when children reach out and grasp - take - an object that has been visually detected. They see something, touch it, embrace it, take it, and bring it to their mouths.
▼

Sees it -

touches it -

embraces it, takes it -

puts it into the mouth!

From such observations of taking, we conclude that children can now perform several important actions.

- Visual performance – They see something on the table, for instance, a toy.
- Anticipatory behavior – The hand moves in the direction of the detected toy and the fingers open up to a grasping position *before* they touch the toy. Children appear to expect that they will be touching the object. This behavior requires previous visual *and* tactile-kinesthetic information about the *same event*. In other words, it requires a coordination between seeing and feeling. In the actual situation, the opening of the hand follows the seeing of the object.
- Holding actions – the fingers touch the object. They move around it and embrace it until the object is held and taken by the whole hand.

The experiences of taking objects are important for becoming familiar with the surroundings. Children see something, and because of previous experiences of touching and holding, become *aware* of it as something that can also be touched. Thus, they perceive it in the true sense of the word.

1.4 The Surroundings Become Familiar

Children are born into a world that is unknown to them. Whatever they touch, see, hear, smell, and taste is unfamiliar. What does that mean? What does it mean to become familiar with the surroundings? As adults, we can hardly imagine a stage of being so unfamiliar with everything.

1.4.1 Unfamiliar – I Jerk Away

In animals we often observe the quick withdrawal from an unknown stimulus. We understand this behavior quite well, since their lives may depend on fast reaction. As humans we, also, withdraw from unknown stimuli. Such a reaction can be elicited through any sensory modality. The following examples illustrate how withdrawing can be a natural reaction:

Touching:
I am in bed and it is dark. Suddenly I feel something on my hand. Something is touching me. I don't know what it is. Automatically I jerk my hand away. Only after withdrawing my hand, do I realize that it was my dog touching me. I am relieved. I may even stretch out my hand to repeat the touching, and this time, I stay with the stimulus. I feel the soft fur of the dog and enjoy the touching.

Seeing:
I am teaching a course at a school for handicapped children. The children are on vacation, so the furniture is piled up on the balconies to make the house cleaning easier. My office opens up to one of the balconies. The weather is beautiful, and I step out of the room onto the balcony. At that moment, I jerk and withdraw. A big shadow right behind the door has caught my peripheral vision. What is it? I quickly turn and look to my side. Ah, a huge doll!
Now I realize what gave me such a fright.

The reaction of jerking or withdrawing can, therefore, be elicited by stimuli which are unknown to us. It can also be elicited when our expectations of the quality of a touched object are contradicted.

We have made a fire to roast hot dogs. D. finds a branch to stick into her hot dog so she can hold it over the fire. With a cry, though, she drops it. Her hand has touched something soft – a caterpillar crawling along the branch.

During the first weeks of life babies are unfamiliar with the stimuli around them. They react with movements involving their whole body.

Such reactions seem quite natural and not abnormal to us. Our interpretation is: *Babies jerk away from unfamiliar stimuli.*

With broader experiences in touching, these reactions decrease. Babies begin to *inhibit* their jerking movements and concentrate on a stimulus for a longer time. The *surrounding things* they touch *become familiar* to them. How does this happen? The next sections of this chapter shall give more information about this.

1.4.2 Searching for New Events

It is fascinating to observe babies' eagerness for new happenings.

Such behavior can be observed over and over again. Babies continually search for new events. When they are successful in taking something new, they can concentrate entirely on that new thing for some time.

T., 7 months, has just explored a strainer next to her on the support. Now she sees something else – something new.
She releases the now familiar strainer.

She embraces the new thing on the support – first with her fingers and then with her mouth.

J., 7 months, is able to grasp a brush. She concentrates entirely on the feeling and even closes her eyes.

Then something else happens. She no longer pays attention to the brush. She investigates the new event by looking at it.

When children notice something new, they seem to focus first on one input modality – feeling *or* seeing *or* hearing. Behavior characteristics of a *modality-specific integration* can also be observed in adults, e.g., when they focus intently on one event. I like to listen to music. If I want to concentrate on a specific part of the music in a concert, I may close my eyes. When I admire a beautiful picture, I am grateful if I don't have to listen to someone else talking (see paragraph 1.3.1 "I See or Hear or Feel").

Only when children become familiar with an event, are they able to both feel and look at the same time and thus integrate the information of two modalities (see paragraph, 1.3.2 "From Feeling to Feeling and Seeing").

Children discover new events every hour of the day. We can observe the same *sequence* each time: seeing, touching, and holding – first with the hand and then with the mouth.

R., 11 months, is sitting on the fresh snow. She is looking at it,

touching it,

grasping it with her hand

and then with her mouth!

Most importantly, getting familiar with our surroundings means touching and exploring what is around us and feeling it with our bodies. The hands have an important role in this interaction. By watching children we can observe something startling.

1.4.3 One Hand – Then Two – And Still a Unity

The sequences of touching and embracing become longer in duration and the accompanying movements develop in coordination and skill. At 7 months, babies already demonstrate harmony in moving their fingers around an object. Children at this age will explore something new primarily with *one* hand (Forman, 1982). After some touching has already been done, the other hand will be used on the new object. The following examples illustrate this:

E., 5 months, guides a ring to his mouth He sucks and bites on it, holding it with one hand. The other hand approaches it in a touching position.

Now he embraces the ring with his second hand.

R., 8 months, is sitting in the grass. Colorful spots are all around her.

She inspects them by looking. Can she take them off? Carefully she reaches out with one hand *while* the other hand *is held in "free space" as if forgotten.*

She has taken off one of the spots and touches it. Now the other hand becomes involved in the task, touching, caressing and moving it!

What a miracle! Here one hand offers a source of information; then the other hand comes along and gives a second source. How do children succeed in bringing together two different sources of information and ultimately construct a unity of knowledge?

Unfortunately, the literature does not tell us about this miracle. We read about dominance – left-handedness or right-handedness, but there is little information regarding the amazing development concerning the *unity of knowledge about our world* that results in spite of having two different hands and feet touching it.

1.4.4 The Multitudinous Ways of Touching and Releasing

When children become familiar with a new event through touch, the variety of touching movements they make is surprising and is an important basis for development. The following example illustrates this aspect:

T., 7 months, is outside lying on a blanket.

While one hand touches a basket cover made of straw, which is on the support, the other one reaches out for a plastic strainer which is also there. Notice how she has to stretch her whole body over the support!

She pulls the strainer closer to her body. Her other hand has released the cover.

And now the free hand touches the strainer...

so that both hands touch and embrace the object on the support.

Then one hand releases the strainer and returns to the cover, releases it,

and goes back to the strainer. Touching it with one hand,

she now attempts to embrace it with both hands.

Next, releasing the strainer, she quickly touches the cover again

but goes right back to the strainer.

Now she is successful at holding the cover and taking it off the support with one hand. The resistances between her body, the support and the sides change drastically.

She embraces the cover with one hand,

touches it with the other hand,

and now with both hands.

After releasing the cover, she returns to the strainer.

Here she finds something on the side of it – a handle – which she can embrace with one hand

and then with both hands.

How many contrasts this child is perceiving with all of those different touching movements! Each time she moves to touch something, her whole body moves on the support. Thus, the body movements continually change the resistance between her body, the sides, and the support. At the same time she perceives some contrasts of surfaces, of shapes, and of temperatures.

Eating situations also provide outstanding instances of touching and being touched.

M., 12 months, likes to eat chocolate cake.

"*I touch and the surroundings touch me – on my fingers, on my hands, in my mouth. Where else will I be touched*"?

D., 11 months, finishes his meal.

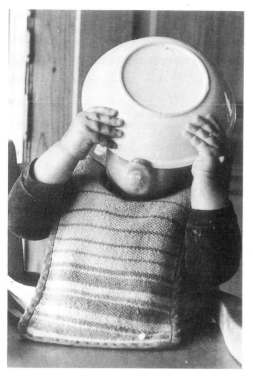

"I embrace my surroundings and they embrace me. First like this; *then like that"!*

To Summarize:

The world becomes a surrounding world for children.

- They are now familiar with changes in resistance which involve the support, the sides, and their own bodies.

- They recognize that there is a world which is all around them. They find security in the niche.

- They discover that they can touch something which is seen by directing their attention to it; they *perceive*. The *surroundings* begin to exist for them.

- They become *familiar* with their surroundings by using a variety of touching and releasing actions, first with one hand and then with both hands.

Children are now ready for another step in development: They *begin to act upon* their surroundings.

2 The Wirklichkeit: Perceiving and Acting Upon

The surroundings have to become *Wirklichkeit*. This requires that children get to know about *causes* and *effects*. They experiment by acting upon the surroundings. Perceiving is always included in this process.

2.1 Cause and Effect

Children discover:
- They move on a support and their movements are stopped when they touch something – cause and effect; they discover that *by moving they can touch* something. They touch something on a support and it moves – cause and effect; they discover that *by touching they can set something in motion*.
- They embrace something on a support, take it off, hold it up, and put it back again – cause and effect; they discover *taking* something *off*.

The following two paragraphs discuss the importance of these discoveries for getting to know the Wirklichkeit.

2.1.1 I Set Things in Motion

When children explore their surroundings through touching and embracing, sometimes they set the things they touch in motion.

At first such movement comes about by chance. Children jerk or at least appear to be surprised by what they have done.

T., 8 months, explores a pan.

She touches it with her hands and then with her mouth.

Now she provokes a change in resistance. Both the pan and T. begin to move. She looks at the moving pan.

Then she succeeds in grasping the handle.

It goes into her mouth.

◂ *Back to the pan! With one hand...*

◂
and then with both hands, she tries to embrace it and causes changes in resistance. When the thing she touches begins to move, her arms and legs move, too. All of this happens on a stable support.

Eliciting movements on the support occurs concurrently with changes in resistance. With growing experience, children get to know that there are regularities between various kinds of touching, and the resulting changes in resistance and moving. They learn about their interrelationships and about the regularity of causes and effects. They begin to experiment systematically.

◂
J., 20 months, is happy when she has the chance to strike the keys of a typewriter.

What concentration it takes to push down on a key in such a precise way that it only gives the amount of force needed to create the correct change in resistance – to make the key move.

2.1.2 I Separate and I Bring Together

We described how children explore their surroundings to find out what can be embraced and taken. Their fingers slide over the support surface, lightly scratching it, in search of something on the support that can be embraced. Or they see something – a spot – on the support and try to touch it and embrace it. Sometimes, though, the spot they have seen can neither be embraced nor taken off the support.

J., 7 months, lies on her stomach on the floor near a small table. The sides of the table are made of pine. One can clearly see the areas leftover from where the tree branches had grown – the knots. These spots are darker than the surrounding wood, and J. tries to grasp them. She tries to grasp them again and again but cannot, of course. During the next few days, though, when she is near that table, she continues to try and take off the dark spots on the sides of the table.

Children show great perseverance in repeatedly trying to embrace something they see. The spots can be touched but may not be grasped or taken off. However, since they cannot predict this possibility by looking at the darker or lighter areas, they must experiment.

It is this way that the child in the example made an important discovery: I can see something – but seeing does not tell me if that spot is part of the support or something independent of it; seeing does not tell me if that something can be embraced and taken off. In order to find out, I must touch what I see and try to embrace it so that I can attempt to take it off. The literature often describes such performance as figure-ground-differentiation (Ayres, 1973) and frequently views it as a visual performance (Koffka, 1963).

When children make this important discovery of taking off, it goes beyond perception. It is a new performance and is more complex than the one described in paragraph 1.3.3 "From Looking to Taking". That performance requires a differentiation between something that moves on a support and its stationary support. Later on children learn that this something that moves on a support can be touched and embraced. Beyond that, children discover that the thing that moves on a support can not only be embraced, but it can also be *taken off* the support and put back on again.

Taking something off the support *changes the surroundings*. That which was on the support at a given location can be removed and is, therefore, no longer at that location. It can be returned to the same place and released there. In this way children learn that they can reverse the effect of taking off. They separate something from the support and bring it back; they put that something together again with the support.

With these kinds of explorations, children experience regularities of cause and effect (in the form of changes in resistance). This is the basis for the "rule of taking off," one of the rules of acting upon.

Children now intently explore their surroundings by applying this rule and asking the question: What can be separated and what belongs together?

R., 10 months, is sitting in the back seat of a car on his mother's lap. During the ride he repeatedly grasps the hair of the woman in front of him. He pulls so hard it hurts her.

One can assume that the child in this example has been looking at the woman's hair. He has also seen the color of her flesh. Since her hair is a different color from her body, he has two different visual impressions. Do the two colored impressions belong together or could he take them apart? He grasped at what he saw – the hair. Is the resistance total? If this is the case, the two colored impressions belong together. If this is not the case, the two colored impressions are separate.

R. pulled as hard as he could but they didn't separate.

This is an example of how during the first months, children need maximum changes in resistance in order to judge if two things belong together or can be separated.

Similarly, children examine glasses, teeth, and even the tongue of those around them. They repeatedly explore to see if such things belong to the body of the person or are separate from it.

How often have you seen children trying to take off someone's glasses? Because they need maximum contrasts, they pull them off with great force.

The following examples are only a small sample of experiences children accumulate by exploring their surroundings and using the rule of taking off.

Over and over again, even when older, children explore the relationships of "being separate or together". The only difference is that the situations become more complex as children advance in age.

D., 7 1/2 months, sees his mother's teeth. "Can I take those off"?

M., 12 months, is eating a piece of melon.

"What can I take off of the melon"? She bites. It is still smooth.

Then she notices some resistance. "Ah! Another taste"!

▲
D., 15 months, sees yellow spots on the floor (onion peels).
"Can I take them off"?

▲▲
J. is 2 years old and wonders about plants.

D. is 2 years, 6 months. ▶
It has snowed during the night, the first snow of the season.
He wonders what it is. "Can I take it off? How thrilling it is to use a tool for that exploration". ▼

The rule of taking off also includes experiences with one's *own body*. Two observations which illustrate this are in the following examples:

M., 22 months, refuses to put on her gloves in spite of the cold winter weather. When her mother puts them on her hands, she immediately tries to take them off again. At the same stage in her life, she offers resistance when the sleeve of her pajama top covers more than half of her hand. She is seemingly afraid that her fingers aren't there if she can't see them.

L., 2 years, is in bed. She pulls down the sleeve of her pajamas and comments, "Look! Gone"! Then she pulls out her fingers and comments, "Look! Not gone! Fixed. Glued on".

The rule of taking off also includes *putting together* again. In the following example, the child is focusing on that step:

E., 2 years, 2 months, has taken off his slippers. Now he tries to put on his shoe.

◀

He holds the shoe near his foot. "Here is where they are supposed to be together – the shoe and the foot"!

"How do I hold the shoe? Where do I touch it"?

His toes search for resistance. "Do they go into the shoe here"?

He presses the shoe on his toes. The resistance becomes stronger. "But where should I put my toes"?
(The shoe is upside down now.)

He begins again. "Maybe it goes like this". The toes touch the shoe. "Do they go in here now"?

During this period, children become more *independent with eating*. The interiorized experiences with the rule of taking off help them to reach this level. What can they take off the plate or out of the dish? How can they do this or that? The following example further illustrates how changes in resistance give important information:

▶

M., 18 months, is sitting on the kitchen stoop with her bowl of ricotta cheese, one of her favorite foods. She puts the bowl on one knee. With the other knee, she creates a resistance on the side. The right hand holds the bowl down on the resistance of the knee that is supporting it and against the knee which is at the side.
Now the bowl is stable. Finally she can begin to eat – to take off – from inside the bowl whatever does not resist movement.

To take off requires holding, lifting up, bringing the object together with a support, and *releasing* it there. The development of the releasing performance is easily observable.

In paragraph 1.4.4 "The Multitudinous Ways of Touching and Releasing," we described the various ways of touching. Children find their way to an object across a support. When they see another object, they also try to reach it across the support. To do this, they usually have to let go of the first object before they can reach for the second one. The changes in resistance elicited by all of these experiences of touching, releasing, and touching again always include the support.

During the earliest stage of experiencing changes in resistance between the support, an object, and the body, releasing is elicited more or less *accidentally*. Only with an increasing variety of touching experiences does it become more conscious and systematic.

Therefore, the *voluntary kind of releasing* requires *previous tactile-kinesthetic experience of touching and embracing objects on a mutual support*. To release an object, children must consider a sequence of changes in resistance which is the reverse of the sequence for taking off.

- *When taking something off a surface, we can feel the changes in resistance between the support, a side (object on the support), and our bodies. First we feel the resistance of the support in contrast to the moving object on that support; then we embrace the object with our hand, feel its resistance, and take it off the support.*
- *When releasing an object, we hold the object in the hand and feel its resistance. Then we touch the support. As soon as we feel the resistance of the support, we press the object against the support with the palm of the hand, and the fingers let the object go. The palm of the hand moves off the object. This is what is meant by releasing an object on the support.*

The taking off experience, including releasing, is connected with the *relationship between a support, an object, and our own bodies either being separate or together.* Through explorations with this kind of taking off, children acquire numerous experiences in finding out if "spots" which they see can be embraced and taken off or if they belong to the surface.

At first, such experiences are only connected with *one* object and its relationship to a support and to our own bodies. With increasing practice, children notice that by taking something off the support, they can meet another resistance, and perhaps put it in motion, too.

Repeated occurrences of such events permit children to discover the regularity of cause and effect. This is the basis for *recognizing relationships between objects which touch each other* in the environment. Children learn the rule that they can take an object and, with that same object, touch another one which is on the same support. By touching it they may even put it into motion. With this discovery they can now begin to apply this rule to explore neighboring relationships in a systematic way.

2.2 I Explore Neighboring Relationships Through Feeling

When objects touch each other, the neighboring relationship that children discover and explore very early is "taking out" and, related to it, "putting in".

2.2.1 I Take out and Put in

This performance is closely related to taking off. While exploring taking off children often come upon the following situation: They see a spot on the support and try to embrace and take it off. During the attempt, their hand may go into something. They experience resistance while sliding along a side until their fingers touch and embrace

J., 2 years, 8 months, explores her little sister's mouth.

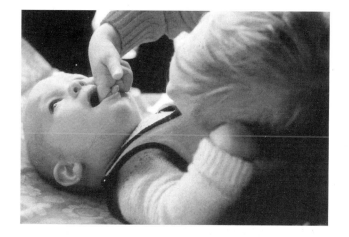

"Can I take something out"? She puts her fingers inside. She sees and feels the tongue moving. "Perhaps I can take it out"!

the spot they had seen. The thing embraced can be taken out along the resistance of the side. In this way they discover taking one thing *out of another.*

J., 9 months, touches a little box in which there is a die. She can feel that there is something in the box and tries to put her hands inside. However, this is quite difficult since the box is small and there is hardly space for her fingers. In spite of this, she tries again and again to put her fingers into the opening in order to grasp what is in there and take it out. After several attempts she is successful. During the next few days we can observe this same behavior many times.

The more developed children are in locomotion, the more extended are their fields to explore taking off and, its related task, taking out. Soon nothing is safe in their surroundings. e.g., books are removed from the shelves they can reach, and if by chance a cupboard door is left open in the kitchen, it can soon look like a disaster area.

In the same way that taking off is related to releasing or bringing back and putting down, taking out is related to *putting in* again, but now the performance is much more difficult.

M., 10 months, likes to have glasses or cups so she can put spoons inside them. She also grasps food from her dish and tries to put it into her mother's mouth while being fed. During this same stage, she puts her spoon into her dish and then into her mouth.

D., 15 months, has taken the onion peels off the floor (see example p. 43).
Can he put the peels back into the plastic bag again?
"Do they go in here?
This is where they were before".
He holds the peels next to the plastic bag.
The bag and the peels belong together.

47

Children explore such relationships in many ways: taking off, putting on, taking out, and putting in. The simple sequences of "object off" and "object back" slowly become more complex events. As other objects are included in the sequence and reciprocal relationships in the surroundings are explored, taking out becomes taking apart, and putting in becomes putting into another.

Children explore with buckets, spoons, cups, cans, and many other things at this stage. Whatever can be taken off and filled into something else – sand, soil, or water – is moved with the hand, the spoon, or another tool. When something is inside, they try to take it out. Children are busy with such experiments for many years.

Experiences like these are the *foundation* for acquiring *knowledge of cause and effect relationships in many different situations and, later on, for forming concepts of verbal and mathematical operations.*

K., 2 years, 2 months, explores her surroundings by putting things into one another,

taking them off in parts, taking them off all together, and putting them back.

J., 2 years, takes her bath. What a magnificent field for explorations!

The next example demonstrates the *spontaneous variations in the repetition of such actions* by a normal child at the age of 2 years, 6 months:

J. and his father are sitting in front of a box with several objects in it. J. takes out an orange juicer and gives it to his father. He also takes out a can and gives that to his father to hold. Next, he takes a stone out of a can and gives it to his father. He points to the can and says, "Put in. No, not put in".

J. then takes out a funnel, puts it upside down over the can and tries to put a little stone into the narrow end of the funnel. He comments, "In". Then he takes a different juicer, one that has a container attached to the bottom to catch the juice. He lifts off the upper part and discovers the empty lower part. He puts a nail into this. Then he realizes there are more nails on the floor and adds them to the one already inside.

He sets the upper part of the juicer on a drinking glass and then removes it. Now he puts the lower part on top of the glass and adds the juicer to that. He pushes the whole thing and everything falls to the floor. He takes them off the floor and starts his construction all over again. One part begins sliding down. He takes the juicer off the glass and now puts an orange on top of the glass.

In the experience of putting something into something else, there may be no resistance to stop the action. For instance, when children try to put a washcloth into a hollow tube, they may push and push along the side resistance of the tube, but may wonder where the bottom is that would stop the washcloth from moving? Instead of meeting resistance at the bottom, the side resistance suddenly ends. Where did the washcloth go? There it is, at the other end of the tube. It has gone *through* the tube because it has no bottom.

The more often children have such opportunities, the more they become conscious of another relationship of neighborhood.

2.2.2 It Goes Through and Then Where Is It?

M., 12 months, discovers a gap between the wooden boards of a bench.

◄
She is holding a piece of paper in her hand.
"What can I do with this paper"?

◄
"I could put it through that gap".
She tries to do it, pushing and pushing.
Suddenly the resistance is gone.
"Where did the paper go"?
Her fingers don't feel the paper anymore. They only feel the resistance at the sides of the opening.

◄
M. looks through the gap and sees the paper she pushed through it on the ground underneath the bench.
She bends down to pick it up.

Children encounter the new relationship of "going through" in a variety of ways. Here are two more examples:

R., 18 months, is playing in the waiting room of the Center. Two different sets of disks are on a table. One set has a hole in the middle and must be put in a row over a bar. The other set doesn't have an opening. The disks in this set must be stacked one on top of the other so that each one fits inside the next larger one. Both kinds of disks are mixed on the table.
R. holds the bar with one hand. With the other hand he takes the disks without really choosing them. He tries to put each disk on the bar. Of course he is only successful with those disks having holes in them. After several attempts he tries to do it more systematically by taking a disk with a hole and then a disk without one. This means he has discovered there are two kinds of disks. He sees those with a hole and those without. He is successful in alternating his choices. However he seems unable to judge in advance which ones he can put over the bar and which ones he cannot. To solve that problem he has to tactually try out each disk.

 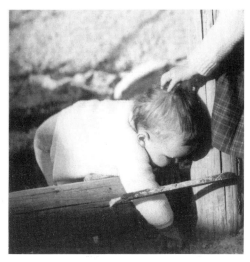

M., 15 months, discovers some little stones on the ground behind a log. He wants to get them. But how? He explores the situation. He touches. Where are the changes in resistance? "Why can't I reach the stones"?

"Ah"! Reaching through, between the log and the iron bar, he finds a restriction of his movements on both sides!

With one hand he touches the ground, and with the other one he embraces the log and feels its stability. He touches and compares. "Here is the stable log; here are the stones which move"!

Now he squeezes his other hand between the log and the iron bar and reaches through to the ground and the stones. Now he can take those stones off the ground.

In these explorations, the object going through something else was always *visible*, and the child could coordinate the tactile information with the visual. However, sometimes an object passes through an opening and then disappears.

The more children collect experiences where they can see *and* feel how an object goes through something else, the more they begin to wonder about those situations in which the *object disappears*. They begin to explore this second condition in a systematic way.

D., 18 months, has discovered an empty aluminum can. He begins to put all kinds of objects into the narrow opening of the can. He especially likes to put in the stones from the road. Some disappear; others are too big to go through.

Walking down the street, he discovers a drain with a grating over it. He also finds different sizes of stones and tries to push them through the holes in the grating. Once again some stones disappear in the drain; others do not.

Therefore, children discover that objects can disappear when they go through something and can no longer be seen. What has happened to them? Where did they go? More about these situations will be discussed in the next section.

2.2.3 It Disappears and I Find It Again

At first objects disappear by chance. Although we cannot see them, perhaps we can still feel them as the child does in the following example:

J., 15 months, was playing with paper clips. One of them disappeared under a piece of paper, but she hasn't noticed because she is busy scribbling on the paper. Suddenly, she feels a little bump and tries several times to take it off. Finally I take the paper away. Now she can see the paper clip and laughs with satisfaction saying, "There"!

After children find objects which have disappeared by chance, they soon actively endeavor to cause the disappearance. For many years this will be a preferred field of exploration. Being under something changes into being embraced by something. At first a child places an object on a paper, not under it. Then they will put the paper around the object, enveloping it and pressing the whole thing together. The strong force applied indicates the maximum change in resistance children must elicit in order to control the success of their actions. At early ages this strong force will be used, for example, while helping to fold napkins or the wash. Children will press the corners together with maximum force.

After numerous experiences with pressing things together, children begin to be able to use smaller changes in resistance. They try to envelope an object by rolling it up in paper. In later years, the combination of pressing it together and rolling it up leads to the complex action of *wrapping it up*.

To make an object disappear by wrapping something around it is a fascinating moment for children, but *unwrapping* is even more so.

M., 19 months, always unwraps packages with great enthusiasm

She receives a package. Something is in the wrapping for she feels the resistance of it inside the paper.

She pulls at the paper until its resistance is overcome. She tries to grasp what is inside the package, to grasp that resistance she can feel.

There is so much paper and it's very long! She stretches out her arm, holding onto the paper as far as she can. Her other hand makes sure that the resisting thing is still inside the wrapper.

Now she can see what was hidden inside. She has found it!

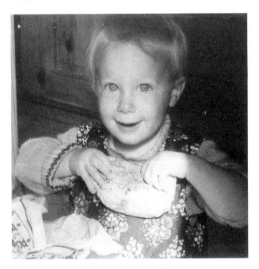

She takes it out. It's a stone!

She laughs and embraces it with both hands. "Isn't it a beautiful stone? I have loved stones for weeks now"!

2.3 The Wirklichkeit Becomes Familiar to Me as It Is

Children's need to act is impressive! They never get tired of touching and exploring neighboring relationships of things in their environment. They search for problems and try to solve them. They hunt for new events and make use of familiar objects in unusual ways. They perceive and act on them. They change actions serving as causes and observe the effects.

2.3.1 I Grasp a Multitude of Causes...

▶

E., 3 years, 4 months, has discovered how to cut paper. He wonders how many ways there are to hold the scissors and still be able to cut.

"Can I hold the scissors sideways pointing forward? How does this feel"?

"Or should I hold them with the points down"?

"Can I lift them off the table and hold them in the air so there is no resistance from the support"?

"Or, to have some resistance, can I put the scissors between my arm and the edge of the table"?

"How should I hold the paper? Straight up"?

"Or flat on the table for a stable support"?

His older brother, D., 5 years, explores cutting by using two pairs of scissors at once.

"Do I hold both of them together"?

"Does the cutting feel the same"?

Later, E., the younger one, helps his mother in the kitchen. They are preparing melons and he wants to take the skin off with the paring knife. This is the effect he wants - taking off.

"But how do I hold the paring knife? Where do I touch the fruit? From above? How do I hold the melon? Where are the changes in resistance"?

Or shall I turn the melon and press it down on the table so that I use the paring knife at he side"?

"The skin is hard. What will happen when I put the paring knife into the soft part of the melon?
Maybe I should put the paring knife away and take out a table knife"? E. gets a table knife.

"How do I hold this knife? How do I touch the melon with it?
Flat like this"?

▲ *"Or do I stab the melon from above? With a lot of force"?*

"No, that's not right!
Perhaps this way"?
E. holds the knife like a pencil.

▶

Then he stabs the knife into the fruit by moving the knife forward?

"No"! He tries it at an angle from above ▼

and moves the knife downward. Now, one piece is cut. The effect is achieved.

How often neighboring relationships were changed in that event – relationships involving the melon, the support and the knife. Each time, resistances changed. "Do I still have the melon in my hand? Or has it slipped away"?

Children become more and more familiar with causes and their effects. The field of experience is wide ranging. They explore whatever event is in their surroundings presuming that something unknown is included.

2.3.2 ...and Effects – Outdoors and Indoors

In the following examples, we will accompany two siblings, M. and J., and also little E. for a while in their daily lives *outdoors* in the garden. These examples will be followed by a common experience *indoors*. In between these two events is an uncommon experience – a visit to the mountains. We will describe the situation and the behavior of the children as precisely as possible and reflect on what kind of new knowledge each particular experience provides them.

M., 18 months, is outside near the stairs that go down to the cellar. The door at the bottom is open and M. has discovered this. Now she tries to go down the stairs.

"How do I get down there"? With one hand, she holds on to the first step; with the other one she holds on to the water main covering. She concentrates on her feet which are touching the next step. With her feet she explores carefully in order to avoid falling down into an empty space.

She touches the next step. Now she can also see where her feet are. There is a stable support. She takes another step, going farther down!

She reaches the open door and enters the cellar. "There are so many things to see – to touch – to hold! Can I also take them away"?

She finds a faded flower with its seeds. She sits down. In this position, it is easier to explore it.
She climbs back up the stairs and continues her exploring.

Again something fascinates her. There are little stones on the ground next to the stairs. "Can I take them off, too"?
To balance her body, she leans on the support with one hand. With the other one she tries to take the stones off the ground.

She is successful! Her right hand holds several of the little stones.
On all fours, she tries to lift herself up.

Now! She balances her body, trying to stand on two feet again.

Before you know it, she sees the next item of interest.
Something long and sparkling catches her eye.
She takes it off with both hands and looks at it. It is a pair of scissors.

She shows me the scissors and tries to open them. She wonders what I am saying.

"How many things I find around me! Here is a shovel. Or is it a spoon"? (It's a trowel for the garden.) She holds the trowel just like she holds her spoon when eating. She tries to stab it into a piece of rope.
"What will happen if I do this"?

J., 3 years, 8 months, is her older sister. She likewise explores her surroundings.

She finds a garden glove, touches it, embraces the big fingers, and is amazed. "How huge they are"!

Next she finds a trowel and a pile of string. She tries to cut the string using the trowel like a knife. "Can I cut the string"?

It doesn't work.
She examines the string, feeling it as she also looks. "No, the string is not apart".

On the contrary!
"What a long string"!
With the string in her hand, she stretches out her arm as far as she can. It goes farther and farther away from her body. "How much farther can I stretch it"?

Such observations emphasize how *new situations* continually arise in the lives of children and each time, children try *to become more familiar* with the new situation. Books could be filled with such descriptions! It is hoped that the adults around them will not be in such a hurry that they make them rush through their explorations! It is a process that cannot be hurried.

In the following example, the sisters are walking in the mountains and have just come upon a patch of freshly fallen snow:

M., 18 months and J., 3 years, 10 months, investigate the situation.

◄

J. slowly and carefully touches this new kind of support with her feet.
M. looks with interest at what her sister is doing. She even bites her tongue because of her excitement.

Now M. has the courage to go near the white stuff. With her fingers, she touches the unknown and takes a little of it off. She feels it and looks at that cold, wet stuff that melts – snow!

Now it goes into her mouth. It is so different from dirt or sand!
M. thinks it over.

Daddy has made a snowball. M. takes it and embraces it – first with her hands and then with her mouth. She sucks and the ball gets smaller. Now there is cold water in her mouth which she can swallow!

Not only outside, but *inside* the home, children continually encounter things which catch their attention and serve as a means to explore causes and effects.

E., 12 months, finds his father's case of drafting tools.

He carefully removes everything from the case. In one hand he holds the instruments; with the other one he searches for a place to put the compass. "How can I put the compass back into the case"?

"There, it fits"! The resistance is now total!

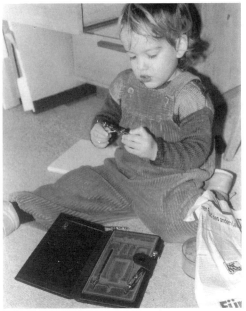

▲
He tries to close the case. He pushes and pushes to feel the change in resistance, but something is in the way. "What? Where"? (The flap of the case has fallen between the top cover and the bottom of the case.)

E. opens the case again. He takes out the compass. He opens and closes the compass. "Will it come apart"?

◄
He puts the instruments back into the case – carefully searching for the correct position.

He makes a new attempt to close the case. Then he notices that the flap is in the wrong place and tries to push it off with his finger. At the same time, he presses down on the top with his other hand and creates a change in resistance.

He does not succeed in closing the case, so he looks for another solution. He spies the newspaper, which is next to him.
E. puts all of the instruments on the newspaper and begins to wrap the paper around them. In this way the instruments now have their place – in the paper!

To Summarize:

It is a long way from *touching the world*, and making sure it exists, to *perceiving our surroundings*.

- Numerous experiences with touching need to occur. By touching, resistances are changed involving our own bodies, the support, and the sides. We can feel safe in the security of the niche.
- Touching leads to being embraced or embracing, to being taken or taking. The world becomes a surrounding world. Tactile-kinesthetic perception is coordinated with vision, hearing, smelling, and tasting.
- The multitudinous and variety of touching experiences involving the relationships between the surroundings and our own bodies are organized by the rules of touching and allow us to become *familiar with the world around us*.

Nevertheless, such a surrounding world is still not the Wirklichkeit!
Living in a Wirklichkeit requires *perceiving* the surroundings, and then beyond that, gaining *knowledge* about *cause and effect relationships*.

- Children notice that an object and support can be separate, and at the same time, belong together. The discovered rule of taking off, and incorporated within it, releasing and bringing together, permit an ordering of cause and effect experiences.

- Children notice that several objects on a support are in relationship to one another. Their fields of experience expand impressively, ordered by the rules of *neighboring relationships.*

- By changing relationships involving objects in their surroundings, their own bodies, and the mutual support, children begin to gain insight into the causes and effects which occur while they are interacting with the surrounding world – they begin to know *what the Wirklichkeit is.*

With the help of a few examples, we have accompanied children along their paths of development. The few examples cited are indicative of innumerable possible observations that can be made. We shall leave the children in this phase of discovering the world and become their companions during another period. We shall refer to this new stage as the "changing of the Wirklichkeit".

B. The Wirklichkeit Can Be Changed

1 The Wirklichkeit as I Want It to Be

The explorations of children when acting on their surrounding world bring about two basic changes: those which remain and those which can be reversed. When we cut paper, the paper is in pieces. We cannot put the pieces together again and make the paper as it was before. When a melon is peeled, we cannot put the peel back on the melon.

In other cases, however, changes can be reversed to restore a former state. For example, when E. took the instruments out of the case and put them back, the Wirklichkeit became *like it was before* the exploration began.

1.1 I Restore the Wirklichkeit

Over a period of time we can observe children as they attempt to reverse changes they caused in their surroundings. The following example illustrates how children systematically explore the *reversal* of changes:

J., 19 months, discovers three coins lying on a book on a table. Quickly she grasps one of them, takes it to the couch across the room from the table, and puts it down there. She goes back to the table, gets the next coin and puts it on the couch as well; she then does the same with the third one. Finally, all three coins are on the couch instead of the table. Later she carries one coin after the other from the couch back to the table. She puts them back exactly where they were before. She repeats this behavior several times, without looking up in between.

Observations like the one in the next example suggest that the child has integrated some features of *the Wirklichkeit as it is* and tries to restore the features if they were changed.

L., 15 months, is in the woods. She picks up a branch, shakes it, and watches how the leaves fall off. Then she drops the branch and lifts up one of the leaves, trying to put it back on the branch.

In this developmental stage, children show a kind of behavior we judge as "orderly".

1.2 I Behave in an Orderly Way

During this period children learn habits. For instance, they learn that shoes belong on the feet or, when not being used, on the shoe rack. Things have their special places. People have their belongings. This belongs to me and that belongs to you.

M., 15 months, is nearby as her Daddy gets ready to leave. Quickly she brings him one shoe and places it next to his foot; then she does the same with the other shoe.

J., 18 months, needs her diaper changed. Afterward, she takes the dirty one and throws it into the wastebasket.

E., 3 years, observes how his brother screws the top of a bottle of water to the wrong bottle, the bottle of cider; later he goes to it and unscrews it.

M., 15 months, often finds a bread crumb or a grain of rice or some other kind of food on the floor. When she does, she picks it up and puts it into her mouth. Such behavior can be observed for several months.

Such "matching" is applied to people, too.

M., 17 months, meets G. on the terrace of their home. Later on she accompanies him to the door and watches as he gets into his car and drives off. She even waves good-bye to him. After a while she goes to the terrace again and appears to be surprised that G. is no longer there.

As children grow, an increasing number of habits are exhibited in their behavior. It seems that they try to *preserve a familiar condition of their Wirklichkeit.*

However, such a quiet period of habit production and orderly behavior, though much appreciated by family members, does not last long. It is soon overtaken by a period of unrest and systematic changes. How does this happen? Is this period of unrest needed?

We will attempt to answer these questions in the next chapter.

2 Events of Daily Living Change the Wirklichkeit

During the period of increasing habit production, children take a more active part in the daily lives of family members. They have learned how to walk and can now follow the others wherever they go: to the kitchen, to the bathroom, to the garden, etc. In these situations, the important thing for children is to not just look at what others are doing, but rather to take an active part in the activities.

2.1 I Help in Daily Living Events

The following examples describe a child taking part in two events of daily living:

Mother has washed the family sheets. D., 15 months, takes them out of the washing machine and puts them into a basket.

"How big that basket is! How far down the pieces go"! He bends over and presses them – feeling them.

"Now it looks quite different". He examines the basket.

"What a huge sheet"! He stretches out his arms until the arm movements are stopped by the sheet.

"Is the machine empty now"? He explores. "Ah! What a secure resistance"! He can feel it with his whole body. There is a gap between the door and the machine. He touches it. "Is there something in it"?

Mother has prepared lunch, and D., 2 years, has been helping to cook. The dirty dishes are on the counter. He wants to wash the dishes.

He is too small to reach the sink but he knows how to solve that problem.

Now, he can see everything and fill the sink with water.

Finally, the interesting explorations can begin – the washing of the dishes.

When performing such events of daily living, situations change continually and present children with opportunities to discover a magnitude of causes and novel effects. The children's past experiences are confirmed, and at the same time, extended. The stock of experiences becomes more flexible. The already familiar rules of touching and acting upon become included in more complex events. Children begin to identify problems arising in the course of daily events and take part in solving them. Thus, through problem solving activities, they become familiar with the changes elicited in the Wirklichkeit.

2.2 I Continue to Perceive and Act Upon

Children zealously want to help both inside and outside the house. This is exemplified by M. and E. in the following examples. M.

is watering the plants and E. is hanging up the wash.

These examples emphasize the close relationship between perceiving and acting upon; they result in changes of the Wirklichkeit:

M., 18 months, starts to water the plants.

The watering can becomes lighter. "Where is the water"? She has to hold it differently again. But how? "Ah! Like this"!

She can, but soon the water stops.
She needs a second hand to help. It becomes difficult!
Both hands are on the handle. "Where can I hold it so that the water will pour out"?

The watering can is full and therefore quite heavy. "Can I hold it by the handle with one hand and have the water flow into the pot of flowers"?

Now she has to bend her entire body. Her knees help too, by creating a side resistance for the can. The can touches the ground, and she tilts it, so that the water flows.

E., 21 months, has found a chair; now he can help his mother hang up the wash.

▲
"How do I hang and fasten the clothes on the clothesline"? This is a problem of neighboring relationships. *Each piece of clothing has to be in touch with the line, and at* the same time, *hang down on both sides. E. tries to do this.*

▶
"Now I should attach it. This is difficult. I must fix the clothespins and hold the piece of wash at the same time! It's a good thing that I have a mouth to help"!

By using the movement of his hands and *mouth, he performs the* causal actions which result in the needed effect*!*
◄

The first clothespin was made of wood, but this one is made of plastic. He tries to do it with his mouth again, but the resistance of this one is different.

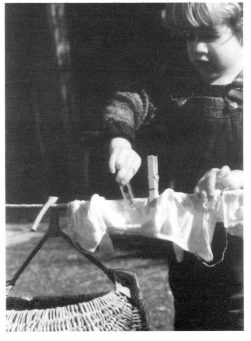

"Maybe I can try it with my hand".
His exploration of the situation is successful.

Again, it's a problem of neighboring relationships.

▲
He holds another piece of wash in his hand. "Where should I put this one"?

"There is still a space between those two".

▶

He balances on a chair to reach that space. His whole body moves. Will he master this situation?

In each one of the examples in this section, the child attacked a *problem* in daily living: D. emptied the washing machine; D. washed the dishes; M. watered the plants; E. hung up the wash.

Solving these problems required various ways of touching, taking off, releasing, and putting together. When performing these events, one must take into account the relationships of neighboring items. The causes and effects which are interrelated must be recognized as such. All of this is directed toward the *goal* of solving a problem. While solving these problems, the Wirklichkeit continually *changes* and demands uninterrupted perception of new situations and adaptation of the ways of acting upon. The washing machine is unloaded; the basket is filled; the dirty dishes are washed; the watering can is emptied; the clothes line is filled. All of these are *problem solving events*.

The more experiences children have with such problem solving events, the more independent they become. In certain situations, they will soon say, "I can do it myself".

2.3 I Can Do It Myself

What pride children have when they can solve problems by themselves within their family living situations. The following examples illustrate this:

J., 2 years, 3 months, takes over several activities in the kitchen.

She chops cabbage on a wooden board. Skillfully she handles the knife and the vegetable.

Then she pushes the cut pieces off the board into the bowl. Attention is given that nothing falls outside the bowl.

Now she cuts the cheese. It is difficult to do! She has to hold it tightly and aim precisely.

The cheese gets smaller and, consequently, more difficult to hold. While doing all this, she cannot think of anything else or look around – she can only concentrate.

D., 5 years, and his brother E., 3 years, 6 months, often help with the dishes. They divide up the work. D. washes the dishes and E. dries them. Each one does a part all by himself.

At other times, D. washes, dries, and puts them away without his brother's help. He performs the task just as he had learned it from his mother. He places the wet washcloth into a cup,

feeling the resistance.

Then he washes the cup on the outside moving all around the sides.

◄

"Is the cup clean now"? He examines it. "Yes". Now he can put it on the rack to dry. But, the daily chore of cleaning the dishes is not finished yet. He has to rub hand cream on his hands just like his mother does.

Opening the tube, he squeezes a bit of the smooth, soft cream on one hand. ▶

▼
Then he rubs both hands until it disappears. His skin feels smooth.

After closing the tube, he puts it away.

Situations in *daily living* always have an *emotional and social* component. It may be strong or weak. It has to be perceived and included in the cause and effect relationships of what is actually happening.

J., 3 years, 10 months, and her sister, M., 19 months, are together in the kitchen. J. has received a package. She puts it on a stool... and begins to unwrap it. M. is standing next to her looking at the stone she loves. She holds the stone in her hands. She had just unwrapped it only moments ago (see example pp. 53-54).

 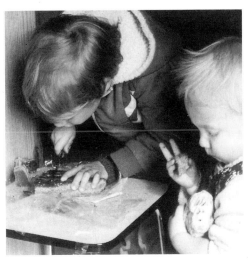

Gingerbread is in J.'s package.
J. begins to divide it with a knife.
"Mmm! Is there also a bite for M."? M. puts her stone on the chair so she can better concentrate on what J. is doing.

"Good"! J. gives her a piece of the gingerbread. Contented, M. picks up her stone again and enjoys looking at it while eating the gingerbread. She is happy!

J. continues to think about the gingerbread as she cuts it. "Who else should receive a piece of my gingerbread"?

"This piece is for Mommy. This one is for Daddy. This one is for me".
M. watches. "Will J. forget someone"?

J. continues to cut.
M. continues to watch. *"Is there another piece for me"?*

"Yes"!
J. breaks off a corner from her own piece and gives it to M.

To Summarize:

Children grow up in the midst of a group of people and in a living situation which allows them to have their first tactile-kinesthetic experiences with the *Wirklichkeit as it is desired by those around them,* including the cognitive, emotional, and social aspects.

Children begin to *restore* the Wirklichkeit of familiar daily living whenever changes occur.

- They use previously acquired knowledge of activities as causes to elicit certain effects.
- They learn habits to help them reverse changes which occur and restore the Wirklichkeit as it is desired by those around them.

In the course of such experiences, difficult and new situations often occur. They demand that children perceive and adjust their actions accordingly.

With the extension of experiences of daily living, the children's knowledge about *the Wirklichkeit as it is* becomes more flexible and adaptable.

This knowledge helps children to make another important discovery: *Daily living events change the Wirklichkeit.*

With this discovery children reach a new stage of development. We will leave them at the threshold of this period. What happens next will be discussed in Part III C, "Tactile-Kinesthetic Experiences with Solving Problems of Daily Living Are Interiorized". Now we will interrupt our reflections and discuss what happens to those children who develop differently from the children we have described thus far.

Part II
Failing in a Wirklichkeit

I'm frustrated – I'm afraid.
Why?

I know quite well
that the world exists around me –
But I can't touch it –
adequately –
nor adequately embrace it and hold it.

That's why I don't know
What the Wirklichkeit is.

How would I be able
to change it?

We have followed the development of normal children for several years. We have been astonished at the variety of ways in which they perceive and take things off the support. We have been pleased with their discoveries and creative actions. We have tried to understand how they are different from us, but our knowledge about their development is still very restricted. Normal children's performances develop so quickly that we have difficulty observing them separately.

We shall now leave normal children and proceed to those children who can only partly perform or those people who are no longer able to perform, daily living situations such as described in Part I of this book. It is difficult to imagine what that must be like.

When people limp we can tell their handicap. Blind people can be recognized by the way they walk or perhaps by their eyes. Children with a cerebral paralysis can be photographed, and people seeing those pictures will pity them. Children with Down's syndrome have a certain physical characteristic and look to their faces.

But how about those who are *perceptually disordered*? They look like other, normal children. If a photograph were taken, we could scarcely see any reflection of their disordered perception. Even if we were to observe their behavior for a short time, they would appear to behave like normal children. We might get the impression that they are not educated very well, and with a bit more discipline, their behavior would surely become more acceptable to society. Or, they seem to be immature for their age. Their parents are often appeased, though, when they are told that their children are merely late in development.

However, if we look closer at the everyday lives of these children, we become alarmed. We begin to realize the extent of their peculiarities. Not just once, but over and over again, they become tense, seem not to comprehend, show fear, and attract attention by behavior that is out of place.

As they grow older, the disorder also affects language development. At school the acquisition of reading, writing, and mathematical operations can become a problem.

When these children are brought to us at the Center for Perceptual Disturbances in St. Gallen, they have already been evaluated and labeled at some other institution or by some one else. They are often described in the following ways: emotionally disordered, retarded in language development, having articulation problems, or having difficulties with syntax, learning disordered, dyslexic, having dyscalculia, autistic, lacking in body tone, spastic, having POS (psychoorganic syndrome), and so on. Most of the time we add our own name to the list of such classifications: *perceptually disordered*. This name refers to something very complex – *something we cannot see*.

Part II of this book is devoted to the difficulties of those who are perceptually disordered.

Adults, too, may be perceptually disordered. They may have passed through a normal childhood, successfully practiced an occupation, and lived a well adjusted life within their social group. Then an accident or a disease occurred which left them with a serious head trauma or brain lesion. Daily living situations became difficult to master and they showed inadequate behavior. The body may be healthy enough to permit them to live a long time, but what sort of life is it with a perceptual disorder? What kind of hope do they have?

In the following sections observations of adult brain damaged persons will be described along with those of perceptually disordered children.

We *forewarn* the reader: Inasmuch as the appearance of normal children is heart-warming to us, the failures of perceptually disordered children and adults is depressing. To make matters worse, we will be immersing you in a world where our understanding is still insufficient.

We will try to describe the failures of those who are perceptually disordered by referring to observations. These observations are selected from thousands of examples collected during 15 years of investigation. We have gathered an enormous number of observations and have continued to compare, analyze, and extend them. At first these observations were fragmentary and did not make sense to us, and as you read along, you may have the same impression.

To aid in your understanding we have grouped the observations.

The people in the environment notice: These observations show weariness, fright, excitement, and/or discouragement of those who are perceptually disordered. Where did the failures begin? Was it the fault of the environment? Of the children? Of the adults? Or was it an interplay of both – the environment and the child or the environment and the adult?

We consider: We shall try to examine the fragmentary observations more explicitly. How do those who are perceptually disordered interact with their environment? How do they behave in the Wirklichkeit? How are their difficulties expressed when they perceive and act upon? What can they do? What can't they do?

We explain: We will look for anything that can bring the fragments together. Will we find any useful, basic explanation?

In pondering these observations for 15 years, it has been a long and laborious search for the "root" that hold the fragments together, for a picture that transforms the senseless into the meaningful. It has been a thorny path – two steps forward and one step back.

This Part II of the book shall take you, the reader, along these wanderings. I hope you understand and persevere.

1 Those Around Them Notice: They Are Deviant

All of those who are perceptually disordered, children and adults, attract attention by their behavior. Depending upon the degree of perceptual disorders and the reaction of the social group, their behavior is judged according to a degree of deviancy.

In this chapter we will mention some of the most frequent statements made by persons who are around those with perceptual disorders. These statements are like the parts of a puzzle – I hold them in my hands not knowing whether they belong together or not.

For the time being, we will leave this situation as it is. It corresponds to reality. Family members notice the peculiarities of their perceptually disordered child or adult member without being able to explain them.

Let's immerse ourselves into the description of the deviances. As an introduction, we will briefly visit a kindergarten for perceptually disordered children. Four of them, all around 5 years of age, are in the "free play" area: B., K., M., and W.

M. declares that she wants to play in the doll's corner and K. can be the father. K. is pulling a wagon with a shovel in it through the room. His wagon keeps running into the table legs.

In the corner of the room he looks around with his mouth opened, and he calls, "I am Daddy"! Then he grabs the shovel, waves it in the air for a short while, and returns to the doll's corner. He shouts, "I am here – finished working". Nobody replies. K. repeats his statement several more times. There is no response.

M. is standing by the stove. She hastily puts a pan on the table, turns and opens the oven door, shuts it, and opens it again. She takes the pan off the table and puts it into the oven. Now she pulls open a drawer, grabs a handful of silverware and throws it on the table. All the while she talks incessantly about prepar-

ing tea and how it is hot and burning. To B. who stoops over the doll's bed, she shouts, "What do you mean, I like to beat children"?

B. stands still caressing her hair with one hand. She touches her sweater and fiddles with it, watching M. all the while. Finally she pats the coverlet in the doll's bed several times, grabs the doll by one arm and yanks it out.

W. rushes around the room, grabs the tracks of a wooden train and throws them on the floor. He seizes a small wooden figure and says, "Hi! Ken driving truck". He throws the figure and a few building blocks, which are next to him, into a truck; pushes the truck towards a broom; takes the broom; lets go of the truck; pounds the broom on the table, the chairs, the window sill, and the shelf. He sees a hammer on the shelf, and so he throws the broom away. Snatching the hammer, he pounds it on the shelf and searches for nails. Other hectic activities follow.

Situations like these are not unusual for the kindergarten teacher. Perceptually disordered children attract attention by acting in one of two ways: They are in constant motion - too hectic, or they rarely move - too quiet.

1.1 They Are Either Too Hectic or Too Quiet

Many of the children who are perceptually disordered bustle from one thing to another without stopping. In the preceding example, M. seems to be extremely busy. Although she behaves as if she might be cooking, her play consists of only very quick and brief actions with the pan, oven, or utensils for setting the table. W. flits around the room like a bumble bee. We cannot distinguish what he is playing. Whatever is near at hand, he uses for the moment. He takes one object and often bangs it against another, like he did with the broom and the hammer.

Releasing is usually done by throwing. He throws the blocks on the truck, throws the broom away as soon as he sees the hammer, and throws the tracks and figure on the truck.

These children who are too *hectic* can hardly sit still. When they have to wait, they move their legs constantly. They hit themselves on the head or fiddle with their hands or wiggle around.

H., 10 years and perceptually disordered, jumps around in the house as soon as he is not occupied. At the same time he appears to be very tense. For instance, he runs to the book shelf, snatches a box of wooden blocks, throws the box on the floor, and laughs and flutters his arms and hands. He snatches at other objects – whatever is within his reach – throws them anywhere, and continues to laugh and flit around.

These kinds of hectic behavior are observable, not only when perceptually disordered children are unoccupied, but also when anything unusual happens. This occurs for example when they find themselves not feeling well or someone has directed them to do something.

D., 11 years and perceptually disordered, becomes restless as soon as he catches a cold or when he has been told to do something. He jumps around, moving his legs in a stiff manner, stamps on the floor, pounds both hands on his chest, and begins to spit. Adding to the confusion, he begins uttering incomprehensible sounds.

Some children who are perceptually disordered appear too *quiet*. They hardly move. They stand and watch what is happening around them. B. in the previously mentioned observations of the kindergarten is an example of a passive child. In that example she does nothing but yank a doll out of its bed. When these children who are quiet are in a group situation, they usually withdraw into a corner. Those who are around them conclude that they are avoiding contact – that they are *loners*.

P., 13 years and perceptually disordered, rarely stays long with the group during lunch. He runs aimlessly back and forth in the hallway or watches the help clean up the kitchen. If he is asked what he wants, he only utters, "Aah"! and leaves the kitchen.

The mother of perceptually disordered G., 11 years, tells us that G. does not want to go to the playground anymore to be with the other children. When sent there anyway, he stands at the edge and watches the others play.

E., 14 years and perceptually disordered, remains in a group situation for only a short time. She sits at the table as long as she is eating; when finished, she immediately withdraws. If the children and the teacher are sitting in a circle and reporting past events or talking about a future one, she will soon ask to be excused. After coming home from school, the first place she goes is under a favorite hazel bush where she can be alone.

In some situations we find these children to be quiet, while at other times they may appear active. Here are two examples:

A., 9 years and perceptually disordered, should be getting ready with the other children as they change to their swimming suits. She does not move, but only watches the others. This differs from situations when she is alone and is able to change quickly into her swimming suit.

L., 5 years and perceptually disordered, attends a program for special education. There she is known as a very shy girl. Even when holding her kindergarten teacher's hand, she walks through the hallway with rigid movements, a bashful smile on her face. During group activities she stands immovable, twisting her sweater and looking embarrassed.
This sort of behavior shown in situations where she encounters unfamiliar people, like in the hallway, or in a group event, is in contrast with her behavior in familiar and highly predictable situations such as the classroom. There she laughs freely and plays in a relaxed manner. At the end of the school year, a video of her classroom activities was shown to the staff. Everyone was astonished at her liveliness.
When the kindergarten teacher is replaced, L. needs three months before she can spontaneously go up to the new teacher and lean against her as the other children do.

Not only do these children have bizarre characteristics of movement, but they express themselves in other ways that are difficult to interpret. Children who are perceptually disordered are usually delayed in beginning to speak. Once they know how to speak, though, it soon becomes incessant.

1.2 They Talk Incessantly

During our visit to the kindergarten we observed M.'s hasty actions and her habit of continuously talking to herself. The following example describes the speaking mannerisms of F:

F., 8 years old and perceptually disordered, is introduced by one of the participants during a workshop. F. has a learning disability and delayed language development.
We encourage him to construct a car. We give him some building blocks and a model of a car made from the same kind of blocks.
F. says, "Car, car, car".
He puts two blocks together, one on top of the other, moves them smoothly back and forth on the table, and says, "Ship".

Then he sticks a peg into one of the holes of the blocks, begins to twist it, holds it at eye level and says, "Photo, photo, photo".
The group of therapists observing him start to laugh. F. continues to twist the peg, saying, "Photo, photo". The therapists continue to laugh. This goes on for 5 to 10 minutes; those who are watching him appear to be amused by his performance. A shiver runs down my spine. What is there to laugh about?

When observing this sort of speech pattern, it is important to think about the content of what is spoken. We soon realize that it is very monotonous. Depending on the person, the content may consist of phrases, requests, complaints, or orders.
In some cases, the spoken phrases obviously relate to the actual situation, as the following example describes:

Mrs. L. has had a stroke and is following a rehabilitation program. During therapy her task is to prepare a cake in the kitchen. She stands in front of the table and looks at the different ingredients which are ready there for the cake. As she begins the task, she suddenly remembers she is not wearing an apron. Instead of asking for one, she declares, "In the kitchen one needs an apron".

In other cases, however, the phrases might be difficult to understand and the content confusing, as in the following examples:

D., 11 years and perceptually disordered, is cutting rhubarb. It squirts. He says, "Because they are squirted it squirts".

L., 5 years and perceptually disordered, wants to boil an egg. She holds the egg in both hands for a long time, turning it again and again. Suddenly she drops it on the floor – splat! L. bends down and says, "Luckily I saw it"! Then she holds the dust pan near the egg on the floor but does not touch it, declaring, "You go on, egg. You go on here". Later on she puts another egg into a pan. The burner is not turned on and there is no water in the pan. L. waits for the water to boil. She says, "When there are bubbles, I will hear it whistle".

M., 6 years and perceptually disordered, comes in for an evaluation. When asked to give a greeting, she says, "Hello". She sits down, and begins to babble like a brook. She talks about her broken school bag, her little sister, etc. Her explanations and descriptions are fragmentary and scarcely comprehensible.

Among many people who are perceptually disordered, children as well as adults, the content of the spoken word often consists only of requests and orders.

B. 7 years and perceptually disordered, appears to be polite and well-educated. She drops a toy and it goes under a table. She does not move. She only says, "Oh, it fell down. Please, pick it up"! People around her scurry to pick it up.

Perhaps this is the way B. hides her inability to bend down. The first of the following examples illustrates her difficulty.

A short while after the previous instance, B. needs to get down from a chair she had climbed up on with someone's help. She stands on the chair, wiggles around, and loses her footing. It seems obvious that she does not know how to plan her movements to climb down.

C., 12 years and perceptually disordered, wears laced shoes in the wintertime. He can hardly untie them, so he usually sits on a bench and orders his classmates to do it for him. Also, in other situations, C. can be seen as a verbose dictator. One day the cook tells the group of children that she is very busy and they must put their dirty dishes on the table outside the kitchen – not inside as they usually do. Shortly afterward, she hears C. talking with an authoritarian voice and then – silence. When her tasks are completed, she opens the door and sees the plates and cups piled up properly. She ponders: C. could never have done that; he must have ordered the others to do it.

As we have observed, adults who are brain damaged complain and request quite a bit, too.

Mr. F., is brain damaged. He is in therapy. When he is asked to perform some movements, he points to his leg and says it is hurting him or that he is sick or that he needs a glass of water.

Mr. K. has had a stroke. When he is asked to do something he answers that he is too tired today or asks the therapist to do this or that for him. He demands often from the other patients that they go and fetch things for him. Even at home he appears to harass his wife, behaving like a dictator.

Is this interpretation accurate?

Mr. K's family told me that he used to make fine chocolate cakes. I mention to him that I had heard about the delicious chocolate cakes he can make. "Oh yes," he beams and answers. "My chocolate cakes are fine". I suggest that he bakes a chocolate cake for tomorrow. He hesitates, "No, not today; I'm tired. I would rather go up to my room".
I try to simplify the situation. Mr. K. is led to the front of a table where all of the utensils and ingredients needed to prepare a chocolate cake are laid out. I wonder whether Mr. K. would really do it. He insists he's tired. He says, "Why don't you make the cake"?
Now we put away all of the utensils and ingredients except those needed for the first step of cake baking. Only the bowl, butter, and a wooden spoon are in front of him. Without saying a word, he takes the butter, puts it into the bowl, and stirs it with the wooden spoon. While he stirs, I put the sugar beside the bowl. Mr. K. notices the sugar, puts down the wooden spoon, and pours the sugar into the bowl, and begins to stir the mixture again. Next, I put the flour beside the bowl. The same thing happens: Mr. K. sees the flour, puts down the wooden spoon, pours the flour into the bowl, and stirs it. With the milk and the chocolate, I follow the same procedure. Each time, Mr. K. immediately carries out the corresponding action. He has become silent. Not even once does he look up from his work or sigh or declare that he is tired, but rather, he works and works until the batter is ready and the cake can be put into the oven.

What happened? We concluded that Mr. K. must have great difficulty with organizing a situation. Usually, everyday living events are complex. Mr. K. cannot meet these kinds of requirements, nor the demands of his therapy. Because he cannot organize these normal circumstances, he cannot decide what has to be done. He recognizes his inadequacy, so he says he is tired or asks somebody nearby to fulfill the demands which are made of him. The same applies to Mr. F. in the first example. His leg hurts or he asks for a glass of water each time he is required to perform some action. If we simplify a situation so that only the necessary items for the next step of the action are in front of these people, we take the task of organizing off their shoulders, and they are then able to decide what has to be done and will perform that step. It is not true that they don't want to do something. It is just that they can only fulfill a demand of the environment if the situation is simple enough. Then they can do it. In another observation of Mr. K, we verified our interpretation.

The next day Mr. K. is to prepare breakfast. Again everything is on the table: milk, eggs, tea, coffee, bread, butter, and so on. Mr. K. says, "Oh, today I am not feeling well. I will come back later. You do it". However, just as soon as the bowl and the eggs are the only items in front of him, he begins to crack one egg after the other into the bowl. When finished, he begins to beat them. I put the milk beside the bowl. He sees the milk and says that he needs water, not milk, because the scrambled eggs would be better with water. So I put water on the table. He takes the pitcher of water, pours the water into the bowl, and goes on beating the mixture. The same thing happens with the salt.
Then I say, "Now, let's scramble the eggs". The stove is behind him, and not in his visual field, so he replies, "No, you do it". We help Mr. K. to turn around so that he is in front of the stove. We place the pan on the burner and put the bowl with the egg mixture beside it. Without any comment, he pours the batter into the pan and stirs it skillfully. The scrambled eggs are ready. Three plates are in a pile on the table. He takes one of them, fills it with scrambled eggs, takes a fork, and begins to eat. He appears to have forgotten about filling the other plates. He has forgotten us as well, even though we are sitting at the table, too.

It becomes apparent that, in a simple situation, Mr. K. can work without complaint. This confirmed the presumption, that for people who are brain damaged, the degree of complexity is critical as to whether they can perform an action or not.
Those who are around perceptually disordered persons report another characteristic behavior and call it aggression. This interpretation can have grave consequences.

1.3 They Are Called Aggressive

D., 10 years and perceptually disordered, goes swimming with his class. The children are in the pool chasing each other. Suddenly D. holds one of his classmates by the throat. She shouts and screams, "Ouch! Ouch"! The teacher tries to remove D.'s hands from the other child's throat; things are tense. She scolds D., "What are you doing? This is terrible behavior! Are you trying to strangle poor G".?
For his punishment she sends him home and calls his mother. She tells her that it looked like he intended to strangle his classmate.

During the next few weeks, D.'s mother receives other complaints. One day, D. is even summoned by the police who reports that on the way home from school, D. willfully broke car antennas. His mother questions him as to why he did such a thing. He replies that he only wanted to know what they were; he insists that he had only touched them.

A., 10 years and perceptually disordered, goes with his class to an indoor swimming pool that he is unfamiliar with. After swimming it is usually no problem to dry his hair with an electric hair dryer. This time, though, he shouts and tries to run away. As the teacher attempts to restrain him, he bites and scratches her.

We encounter the interpretation of aggressive behavior, not only with perceptually disordered children, but also with brain damaged adults. Consider the following examples:

R., 20 years, is suffering from a head trauma. In therapy he is asked to squeeze a lemon. On the table in front of him are a lemon, a knife, and a two-piece juicer. He hesitates for a moment; then suddenly, with a violent gesture, he knocks everything off the table. A panic reaction? Again I put a lemon and tools in front of him on the table and repeat the directive to squeeze the lemon. He sits there and seems to think about it for a long time. Eventually, he takes the lemon and puts it into the bottom half of the lemon juicer, a container that is similar to a bowl. Then he puts the top half of the juicer over it, like a hat.

Ch., 19 years is gravely brain damaged due to a traffic accident. I am told that in therapy he shouts and behaves aggressively. Therapists and nurses are afraid of him and refuse to go on working with him.
I ask if he always shouts when they work with him during therapy or if there are moments when he does not? Since no one can answer this question, I propose that we collect observations and look for a pattern to his shouting.

I accompany the therapist and Ch. to swimming therapy. Ch. floats in the water supported by the therapist's hands. She explains to Ch., "Now I will lift your feet". She lifts Ch.'s feet. Nothing happens; he remains quiet. "Now I will lift your head". Again nothing! "Now I will turn you on your stomach". The therapist turns Ch. from floating on his back to floating on his stomach. Still Ch. stays calm. Then the therapist declares, "Now we will swim". Immediately Ch. emits a shout with such intensity that it sets one's teeth on edge. It resounds throughout the whole swimming pool area.

What happened? We analyzed the requirements of the different situations when Ch. was in the water – those in which Ch. stayed quiet and the one in which he shouted. We noticed that the comments, "Now I will lift your feet,... your head,... and now I will turn you on your stomach," indicated that the therapist would do the action and not Ch. The declaration, "Now we will swim," was different. Probably Ch. realized that he was required to swim, but he also knew he was not able to do so anymore. He had been an excellent swimmer before his accident; in competitions he had proven to be a medal winner, and now he finds himself in such poor condition. What other choice did he have but to shout? We were right with this interpretation, as was proved in future therapy situations. The therapists now paid particular attention as to how the directions were formulated. Ch. no longer shouted in such controlled situations. Later, Ch.'s progress further confirmed our interpretation of the problem. He no longer panicked but became able to articulate his feelings and say "I am afraid".

The fact that people around those who are perceptually disordered interpret their bizarre behaviors, such as, biting, hitting, and so on, as aggressiveness, can have severe consequences for the children and adults concerned. This is exemplified in the following tragic account:

L., 10 years and perceptually disordered, is brought to me for an evaluation. I am told that L. had entered a residential school 3 weeks earlier. After 2 weeks in the school, she was dismissed. The reason given was that L. had become so aggressive that she endangered her classmates. They had even found it necessary to put her into a straitjacket.
A co-worker begins with L.'s evaluation and works with her for a quarter of an hour. Then they both come to my office. I am sitting at the desk and L. is standing beside me. I quietly greet her, calmly extending my hand to her. L. is also quiet, looks at me, and gives

me her hand. She seems to be a friendly child and smiles at me.

I turn to my co-worker standing beside L. She begins to narrate what they did. She is acting very calmly. Then, while speaking she suddenly makes a quick movement - unexpected and intense. This is sufficient enough to evoke strong tension in L. who suddenly bites my co-worker's pullover and entangles her fingers in the co-worker's hair.

I gently stroke L.'s rigid body - once, twice, several times. I also stroke her hand and clenched fingers until I feel the tension diminishing. Now I succeed in releasing her fingers. I take them and softly stroke them over my co-worker's hair - once, twice. Attention appears on L.'s face. Yes, it seems to brighten. It is as if, for the first time, L. experiences what it feels like to softly stroke someone's hair.

After this experience it became clear to me what must have happened at her school. If L. reacted with such an excess of tension because of a sudden movement in the otherwise calm atmosphere of my office, it was only to be expected that she would react that way in an environment like the one present in the new residential school. There were no normal children in that school; each child had a problem. In addition, the new environment was unfamiliar to her. Thus, the sources of tension about her were unpredictable. Was it not to be expected that L. would overreact and not be able to control her tenseness because of the continuous and even increasing tension around her? Those in her new environment added to the general tension by judging her reactions and fearing that she could hurt them. They began to tell her how to behave; they came running toward her whenever she panicked and attempted to prevent her from biting someone - thus increasing the problem; and finally, they came with the straitjacket. How terrible to imagine!

Unfortunately this is not an isolated case. Often, instead of a straitjacket, medicines are given to the child or adult who is perceptually disordered with the hope that the drugs will have a calming effect. The problem with the use of this type of therapy is, the environment doesn't need to change.

In example after example, the blame is placed on the ones who are perceptually disordered. They are judged as being aggressive, self-destructive, etc. Is this justifiable?

In the next paragraph, we consider another interpretation of the behavior exhibited by those who are perceptually disordered: ill-mannered. This interpretation isn't as harsh, but nevertheless, it is a difficult label for family members, especially the parents, to accept: it wears them down.

1.4 They Are Labeled Ill-Mannered

You cannot tell the impairment of children who are perceptually disordered by their appearances. They look like other children, but since they do not behave like normal children, they are frequently judged as *ill-mannered*. As a result, parents often don't travel by bus or train or go shopping with their perceptually disordered child. On the bus, for instance, perceptually disordered children may sit down as soon as there is an empty seat. They really have trouble keeping their balance in a jolting bus. They use their whole capacity to keep control of their own problems. They have none left to look around and be polite to older people. Similar behavior can be observed when perceptually disordered children go up or down a flight of stairs; they need to hold on to the railing. In St. Gallen, next to the special school for children who are perceptually disordered, there is such a flight of stairs leading from a senior citizens' home down to the center of the town. The talk is that the children of that school are impolite and will not make way for the older people who need to hold on to the railing, too. Are the children being *impolite*?

M., 16 years and perceptually disordered, goes shopping; his teacher is with him. For some time now he has known the value of money. Before they go to the cashier to check out, M. counts his money to see that he has enough to pay for what he has chosen. Scarcely finished, he rushes to the cashier. She is busy entering into the cash register the purchase amounts of another customer. M. bends down and holds his money under the cashier's eyes, exclaiming, "Look, I already counted it all up; it is the exact amount". He doesn't seem to notice that the other customer has not yet paid for her purchases. This kind of behavior often occurs in shopping situations.

T., 14 years and perceptually disordered, goes shopping in town. Several tourist buses are standing at the edge of the sidewalk. The tourists are getting off and are blocking the way. T. doesn't look to see if he can get by on either the right or left of them, but instead pushes through the crowd, clearing his way by using his elbows. People stare at him and scold him for such rudeness.

S., 5 years and perceptually disordered, spends her first day in a kindergarten group. She has been helping with the cooking and is carrying a basin of water to the kitchen sink. I happen to come along. Splash! She pours the water on me and my clothing. She laughs and asks, "Are you all wet now"? Weeks pass and a similar situation occurs. Again, I meet S. in the hall carrying a basin full of water, and before I can move out of the way, the water is splashing on me and my clothes. S. says, "Didn't I splash water on you once before"?

Is S. really so mischievous?

Other labels characteristic of children who are perceptually disordered are *liar* or *thief*. I received an unfavorable report about R., 8 years and perceptually disordered. He would often lie and it was, of course, considered a naughty thing to do. I asked for more information and was given the following report:

R. came home from school the other day all excited and said that during gym class he was pulled up to the ceiling. Further inquiries indicated that he had swung on two rings and was therefore only a little way off the floor for a few moments.

Was he lying?

The next example shows the *collecting* behavior exhibited by many children who are perceptually disordered. This is often interpreted as stealing. These children are drawn to very specific objects, for example, felt-tip pens, adhesive tape, or plastic bags. The orientation can increase and become a fixation; collecting of certain objects then appears as compulsive behavior.

M., 14 years and perceptually disordered, is fond of keys. He confiscates them wherever he finds them. One evening, a worker at a garage in the neighborhood of the school comes in all excited and says that one of his client's had come to pick up his car but could not because the ignition key was missing. We check the pockets of M.'s trousers and find the ignition key there.

Was M. stealing?

Often, those who are perceptually disordered show peculiar behavior at the dinner table. This only reinforces the interpretation that they are ill-mannered. Many perceptually disordered children and adults also have flighty movements. When pouring a drink they spill it because they don't seem to know when to stop, or they don't seem to notice when pieces of food fall on the floor or on their laps. They eat quickly and stuff too much food into their mouths. They cannot chew correctly; they just move their jaws up and down. The tongue lies flat and sluggish in the mouth. True chewing may be absent for a long time. Licking remains difficult or impossible for years. Whenever they are given an ice cream cone, they bite it instead of licking it. Sometimes they have to use their fingers to remove food caught between their teeth and cheek. Often they have difficulty managing a fork and knife, and so consequently, they use their fingers. Some of them cannot bite things off very well, and as a result they give away apples or other food too difficult for them to bite. They also dribble or drool as they eat.

Another problem is toilet training. It takes a long time for many of those who are perceptually disordered to become toilet trained; the length of learning time is directly connected to the extent of their perceptual disorder.

To Summarize:

Those who are perceptually disordered are seriously troubled in their everyday behavior. People around them label their inadequate behavior in different ways. We mentioned some of the most frequent interpretations and questioned them, but how to explain their behavior in another way?
To find better explanations, we collected a great number of additional observations. These observations reveal situations in which children and adults who are perceptually disordered are able to perform, but they also show times when their performances break down. Descriptions of what we observed are in the next chapter.

2 We Observe: They Have It and Yet They Don't Have It

In Part I of the book we described the rules which children who are normal apply in the course of their development. There are rules which are related to touching and others which are related to acting upon. Experiences with those rules help children get to know about the Wirklichkeit and how to better adjust to it.
But what about children and adults who are perceptually disordered? At the beginning of this chapter we shall describe how they apply the rules of touching, and after that, how they handle the rules of acting upon.

2.1 They Know About the Rules of Touching, but Where Is the World Around Them?

Touching is a behavior which is observable. We can touch what is around us with our bodies – legs, hands, trunk, head. Touching is a part of our everyday activities. These activities are therefore a rich resource for observing the touching behavior of children and adults who are perceptually disordered. We will describe three aspects of the observations we have made. First we will describe the reactions which are elicited when children and adults who are perceptually disordered touch something; then we will attend to their rules of touching; and finally we will discuss the kind of movements they make while touching.

2.1.1 They Withdraw from Touching, Become Tense and Look Away

The first group of observations describes the way children who are perceptually disordered try to *withdraw* from touching or being touched. Children with normal development also try to withdraw from being touched, but children with a perceptual disorder do it more frequently and become tense, as well.

S., 7 years and perceptually disordered, rebuffs her mother when she touches her and says, "Don't touch me. B. does that all the time in kindergarten".

K., 7 years and perceptually disordered, allows herself to be guided in a task during therapy, but after they are finished, it is quite different. While she and the teacher are walking out of the room, the teacher puts her arm around K.'s shoulder but K. shrugs her off.

G., 5 years and perceptually disordered, has been receiving special massage therapy. After two years, the therapist admits that there is no effect because G. becomes more tense instead of becoming more relaxed. This reaction is quite different from the one exhibited by children who have cerebral palsy. They become more relaxed by massage.

Often we can observe that a child who is perceptually disordered jerks when touching something, especially when it is soft, moist, slippery, scratchy, rough, etc.

D., 11 years and perceptually disordered, is to grate cheese for lunch. As soon as he touches the cheese, he starts complaining, "No, disappear, I didn't want to be here". He is asked to get butter from the refrigerator, and as he opens the refrigerator door, he starts grumbling, "Didn't want". He holds one hand in front of his mouth and begins to drool. In spite of it all, he slowly grasps the butter but his fingers are quite tense. He is helped in grating the cheese; then his fingers are rubbed over the grater to take off the cheese. As soon as D. touches the rough surface of the grater, he jerks his hands off and hits himself on the head.

R., 10 years and perceptually disordered, is guided by his teacher in cutting a melon. Afterward, they touch the pulp to take off the seeds. For a moment, R. allows this to be done and is quite attentive. Suddenly he pulls his hands away and looks for a cloth to wipe off his fingers. Spontaneously he reaches into the melon pulp another time, but after a moment, he begins again to shout and wipe off his hands.

M., 16 years and perceptually disordered, is preparing fish fillets. He lays the fillets in a row on the table and then runs to the sink to wash his hands. He comes back to the fillets, puts one of them into the egg mixture, and again runs to wash his hands. He works in that sequence with each fillet.

When children who are perceptually disordered touch something moist, it results in an increase of *tension* and often leads to strange movements of their fingers as illustrated in the following example:

G., 12 years and perceptually disordered, is helping in the kitchen. He is mashing bananas.

He only uses his little finger and ring finger and spreads out the other fingers.

He clenches his hand and uses it to crush the banana.

With great effort he presses the banana against the support with the tips of his fingers.

He raises his arm slightly, thus increasing the pressure of his finger tips on the banana and the table.

His fingers are sticky and the banana pulp dries on them.

His fingers are still tense.

Sometimes children don't avoid touching but *look away* instead.

D., 11 years and perceptually disordered, is filling a tomato with tuna fish.

With one finger tip he touches the soft, slippery tuna on the fork and tries to push it off into the tomato. His other fingers are spread apart.

The teacher is guiding D. Now his whole finger, not just the tip of the finger, touches the moist tuna, and he looks away.

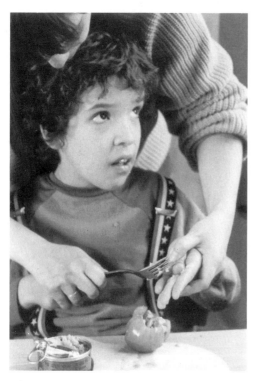
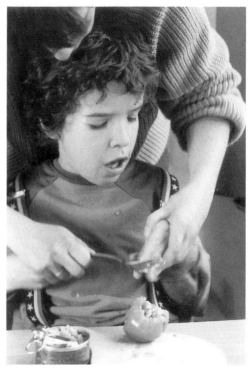

▲▲
"Ugh"! He touches the tuna with all of his fingers. His eyes turn upward. His whole body becomes tense.

▲
Now the tuna is taken off the fork. All of his fingers are involved. He closes his eyes and starts complaining.

◀
Another fork filled with tuna! He holds the can tightly, since this is a more familiar touch; he can now look at what he is doing.

The tuna is on the fork. He looks at it without touching it. Now he smiles – does he recognize the visual information?

He continues his work. Gradually his finger movements become more relaxed. He tightly presses the tuna into the tomato until he feels total resistance – with two fingers. Does he gradually become more familiar with what he is touching?

An especially unique characteristic of children who are perceptually disordered occurs when they are *walking*. They look somewhere else, particularly when they begin to stumble over obstacles. People around them give them directives and say things like, "You have to watch where you're going"!

D. from the last example, is walking down a steep slope, trying to reach a nearby flight of stairs.

He hesitates before starting downhill. ▶

101

He closes his eyes. . .

and begins walking with rigid movements.

He reaches the fence and holds on to it with all his might.

Cautiously he goes step by step wondering if he will reach the stairs.

He feels the wood of the stairs under his foot; now he can let go the fence, his side resistance.

Cautiously he puts his foot on one of the stairs, feeling the resistance of that basic support.

C., 11 years and perceptually disordered, is hiking along a narrow, uneven path. It rained heavily the day before and the ground is now soft and slippery. He falls down over and over again. The teacher takes him by the hand. She notices that C. does not look where he puts his feet; he looks into the air and often calls out something to the other children who are walking in front of him. Again and again, he steps just where it is most uneven. Naturally he stumbles. He is advised to watch the path more carefully and not to talk so much. It doesn't help.

When those who are perceptually disordered do not look at what their feet are doing, it is often quite annoying to the people around them. The more people insist that they should watch out, the less they are able to do it.

In the next few paragraphs we will again refer to the different ways of touching. Whenever we touch, we are searching for regularities. We apply rules. Are children and adults who are perceptually disordered different from those with normal development? Do they know about the rules of touching?

2.1.2 They Know About the Rules of the Stable Support and the Side

Even those of us with normal perception need to check the support and side resistances over and over again. Such behavior was described in Part I of this book. How do children and adults who are perceptually disordered behave in this respect?

The first examples describe observations pertaining to the *rule of the stable support*.

P., 15 years and perceptually disordered, must always make sure of the stability of his support.

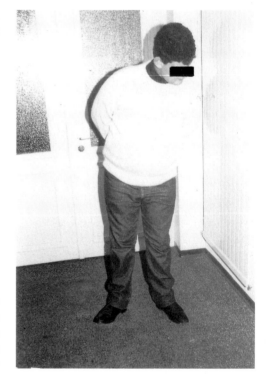

▶

After therapy, he is often required to wait in the hallway while his parents discuss some questions they have with the therapist. While standing there, P. rocks back and forth lifting

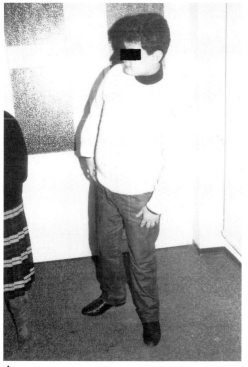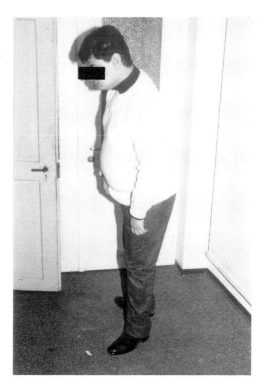

▲
one leg and putting it back on the floor and then lifting the other leg and putting it back on the floor. And so, on it goes.

Those around him interpret this behavior as nervousness. He does stand still for a moment if his mother or teacher asks it of him, but after awhile he begins the rocking again. He behaves in a similar way when he is waiting in a sitting position. He immediately begins to rock the upper part of his body back and forth.

Why does he do this? Does it have something to do with the support? We observed P. as he went up and down a set of unfamiliar stairs.

▶

He holds on to the railing and the wall with both hands. Cautiously one foot glides over the step, checking its stability all the while.

105

He does not remove one foot from the step until the second foot feels the support of the next step below. This is the way he goes up and then down.

His search for resistance information is especially evident when he goes down the stairs.

P. slides one foot forward on the step, keeping contact all the while, until he reaches the edge. Then he lowers his foot, keeping his heel along the side of the step, until he feels the resistance of the support of the next step. Now he sets his foot down heavily and begins the same procedure with the other foot: sliding forward on the step out to the edge. . . .

He stands on the step with both feet.

▲
The same kind of moving and searching for resistance begins. . . .

He feels for the edge of the step and pushes his foot downward, heel first, along the edge.

▶

He keeps his heel in contact with the back of the step. . .

▼

going down, until he feels the resistance of the support.

Now the upper part of his body can lean forward – the new support is stable.

 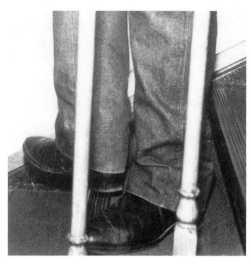

The second foot comes forward, until both feet are standing on the same step, and...

then the movement and the search for information begins again.

This behavior is typical of a small, normally developing child going down a flight of stairs, but P. is considerably older. Why does he need so much resistance?

A similar need can be observed in other children and adults who are perceptually disordered. Recall the observation of R., the 8-year-old boy who "told a lie" about being pulled up to the ceiling on rings (see paragraph 1.4 "They Are Ill-Mannered"). Here is an additional observation depicting R. It describes him searching for the resistance of a support.

R., 8 years and perceptually disordered, is at school where his class is performing a dance. He holds on to the teacher who leads him around the room. He moves in rhythm, but his feet never leave the floor. He is, therefore, in constant contact with the support; he glides over the floor.

A patient is referred to me for observation.

Mrs. R. is brain damaged. According to the report by both a physician and a therapist, she talks continuously, but one can rarely understand what she wants to express. She repeats the same utterances. Sometimes a flood of words is interrupted. We gather observations of Mrs. R. during a therapy session. The therapist is painting with her, guiding her arms and hands. We are able to note the following: The brush is being moved through the air – Mrs. R. is talking constantly. The brush is in the paint container, touching the bottom – Mrs. R. stops talking. The arm is moved through the air – Mrs. R. talks again. Her hand holding the brush touches the piece of paper on the table – Mrs. R. stops talking.

What is happening? Mrs. R. talks as long as her arm is being moved through free space – the air. This kind of arm movement scarcely meets any resistance. She changes her behavior and is silent, however, when her arm meets any solid resistance. This happens when she touches the bottom of the paint container and again when she touches the table top. The patient seems to get no new information from the movement through the air – and she talks. But when there is a *maximum change in resistance*, she seems to be able to perceive it, and she is silent.

Those who are perceptually disordered, both children and adults, often search for strong *side resistances* in addition to exploring the stability of the support.

M., 15 years and perceptually disordered, has a peculiar sleeping habit. He always sleeps at the top of the bed, turns towards the wall, and presses his head against the headboard.

R., 9 years and perceptually disordered, becomes panicky as soon as he can no longer hold on to the side of the swimming pool. He also goes down the pool steps by clinging to the railing and then follows the edge of the pool. He rejects the help of his teacher, even though the teacher promises to hold him if he would let go of the edge.

M., who is perceptually disordered, has shown the following behavior for several years (from age 13 to 16): When he becomes tense, he jumps up, leans against the wall with his hips, and rocks heavily back and forth with the upper part of his body.

C., 12 years and perceptually disordered, still needs to lean against the table when he stands to do some work. The following pictures show him during different actions, and in each case, he needs lateral resistance.

A., 11 years and perceptually disordered, is trying to remove the curtains from a window. First she kneels on a stool. In order to stand up from this kneeling position, she holds on to the edge of the table. She tries to lift one leg. In doing so, she pushes against the wall and moves the table away. Now she is left hanging between the stool and the table and doesn't know what to do.

K., 10 years and perceptually disordered, goes to a gymnasium that is unfamiliar to her. She hesitates before entering; then she goes in but only moves along the walls. During exercises, the instructor must take her by the hand. Her movements look rigid. Even after the group has used the gymnasium for several weeks, K. is not able to run freely but continues to move rigidly.

K., 11 years and perceptually disordered, is supposed to go to the top floor of a school building which is unfamiliar to her. It has huge hallways and a long flight of winding stairs leading to the upper floor. As one goes up the stairs, there is an unobstructed view downward. K. refuses to climb up the stairs even if we try to go along the wall with her. She sits down on the first step and is very tense.

When K. must pick up something from the floor, she is not able to bend down and squats instead. When she was about 4 years old, she was unable to step over a rain gutter that was only 10 cm wide in front of the door to her home.

E., 11 years and perceptually disordered, is supposed to water the plants which are hanging from the ceiling.

While climbing on a chair, she holds onto a table as well as the chair.

Now she is ready and picks up a glass full of water that is on the table.

She tries to straighten her body to reach the plant. But where is a side resistance to hold on to?

▶

She assumes a crouched position in order to search for some additional stable side resistance.

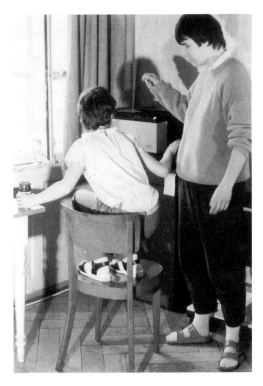

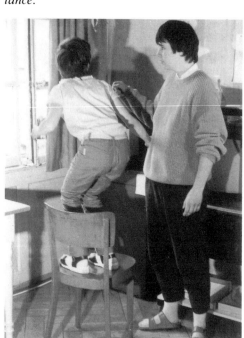

There – she seizes the sweater of the person next to her and tries to straighten her body.

She has reached the plant. She can water it. How lucky for her that the stable side resistance is still there!

In Part I of this book, "Living in a Wirklichkeit," we considered the effect that touching and the resulting experiences of changes in resistance have in helping us to realize that the world exists. Without this knowledge, we would not feel secure.

An outstanding example of a situation in which we can experience changes in resistance is the *niche*. Observations of those who are perceptually disordered indicate that they, too, apply the rule of the niche, and this is often done in an intense way.

Mrs. L. has suffered serious brain damage and is in a rehabilitation center. When she retires to her room she wheels her chair into a niche which is created by the corner of her room and the bed. This position allows her to feel the resistance of the bed on her right and the resistance of the walls on her back and left side. In addition, she pulls the table close to her body so she can touch it with her knees and arms. Now she feels secure.

Feeling secure means having something solid around your body. Touching something that is solid means being able to hold onto it. Holding on to it provides security. To experience such a situation, it is essential for those who have a severe perceptual disorder to elicit maximum changes in resistance. Only when this is possible do they feel secure. Mrs. L.'s behavior is an emphatic illustration of this point.

We conclude: The niche offers the possibility of experiencing maximum changes in resistance through touching or being touched. This permits a person to become conscious of the existence of a stable world and provides, therefore, a sense of security.

This offers an explanation as to why perceptually disordered persons often withdraw into niches.

When by herself, K., 11 years and perceptually disordered, builds a niche in the corner of a room by setting up several chairs as boundaries around herself. She settles into that niche with her toys.

E., 15 years and perceptually disordered, rarely stays in a group when no adult is present. During these moments she withdraws into a shelter she has constructed in the schoolroom.

Further observations of those who are perceptually disordered and their application of touching rules emphasize the importance of the changes in resistance associated with *human contact*. Many perceptually disordered children try to touch the person with whom they are in contact.

C., 12 years and perceptually disordered, is very talkative. Whenever he is talking with someone, he touches him or her. Thus, situations like the following can occur: The teacher is walking with C. through the hallway and down the stairs. C. talks about his activities and asks what he will have to do next. Instead of continuing down the stairs, he repeatedly turns toward the teacher, puts his arm around her neck, strokes her shoulders with the palm of his hand and holds on to the collar of her blouse.

K., 11 years and perceptually disordered, wants to hold the hand of the adult who is accompanying a group of children who are going for a walk. If K. is asked to walk by herself she does so, but stays very close to the adult. In this way she is able to quickly touch the person whenever she feels the need.

E., 14 years and perceptually disordered, is waiting with her mother at the station – her class is going to a camp. E. hangs on to her mother's arm. The teacher arrives. E. lets go of her mother's arm and immediately goes to hang on to her teacher's arm. She gets on the train, still hanging on to her teacher's arm, and stays that way until all the other children are seated. This kind of behavior – hanging on the arm of a familiar person – is observable when she is in an unfamiliar situation.

Sometimes the touching is exerted with such force that it can be painful for the other person involved.

B., 16 years and perceptually disordered, is tall. We try to open a closet together but because of her height I stay at her side and guide her hands. B. turns her face toward mine, smiles at me, and unexpectedly puts her arm around my neck. She squeezes me so hard that I can barely breathe or break free from her grip. B. laughs in her deep, rough voice and bites one of the fingers of her other hand.

K., 10 years and perceptually disordered, is standing close to me. She strokes my hair with an open hand. Suddenly she grabs my hair and pulls it; she opens her hand but pulls it again.

S., 13 years, is perceptually disordered. As soon as he sees a small child, he waves his arms up and down, goes right up to the child, and raps the child on the head with his stretched out fingers. This same kind of behavior can also be observed when he meets visitors. They do not expect such an "attack" and are quite surprised.

Children who are perceptually disordered avoid contact in situations where there are hardly any changes in resistance. An impressive illustration of this is given in the following example where a child who is with her physical therapist refuses to exercise on an unstable ball but does not avoid contact when in a narrow setting.

B., 5 years and perceptually disordered, has delayed motor development and receives physical therapy. The therapist soon notices that it is difficult to work with him. He cries when he has to lean on the big ball. With time the situation becomes worse; he starts to cry as soon as the therapist approaches him. He was labeled a tactile defensive *child who avoided, and even refused, contact with others. He did not talk even though he understood commands.*

B. comes for an evaluation. It is a teaching situation and about 20 participants are seated in the big room where I intend to work with B.
We have to construct a makeshift work table for the child and me by utilizing two boxes and a wooden board. The space between the boxes is so narrow that it will be hardly possible to place four legs there.
I wonder how I will be able to work in such close physical contact with a child who is described as avoiding contact. I put some objects on the table which I hope will remind him of some of the activities his mother performs in the kitchen. For his activity I have chosen the preparation of a fruit salad.
I approach him and briefly say, "Look! I have some work for you". Then I take his hands, sit down with him at the improvised table, and immediately begin guiding his hands for the first step of the event. This happens so quickly that I can only bring one of my legs into the space under the table. Finally, while continuing the guiding, I am able to put my second leg under the table. Can you imagine a stronger contact than what we are experiencing? Somehow there are four legs under a table where there is only room for three. I am

sitting on the chair pressing his body against mine so that I can guide his hands with my hands. How is he behaving? He doesn't even flinch as we encounter such strong contact. He seems unaware that I am guiding his hands. The whole time he focuses his entire attention on the task we are performing. His behavior surprises the course participants. It seems to be in such a contrast to what they were led to expect. B.'s mother can hardly believe that she is seeing her boy in such close contact with an unfamiliar person – and working attentively, too.

What is happening? Let us compare the two situations! In the enclosed sitting space, B. can feel stable resistances all around the body – below from the support, on one side from the box, and on the other side from my body. This enclosure is so compact that each movement of his body experiences total resistance. The situation during physical therapy was quite different. There the therapist worked with B. in a big room with almost no furniture. Where could he find the resistance from his surroundings in such a setting? In addition, when practicing on the big ball he experienced instability. All this created a very insecure situation for him, and he reacted with panic.

The examples in the paragraph emphasize how those who are perceptually disordered, children and adults, search for changes in resistance. They know the rules of touching and try to apply them; in using the rules they elicit strong changes in resistance.

In the next paragraph we will discuss some movement patterns that can be seen when children who are perceptually disordered touch their surroundings.

2.1.3 They Have Two Hands but Often Use Only One...

In Part I, "Living in a Wirklichkeit", we discussed how children who are normal in development use their hands. They begin by utilizing one hand at a time as they touch their surroundings – one time it is the right hand; another time it is the left one. This period of *one-handedness* leads into a period of *two-handedness*. To be able to utilize two hands at the same time requires a complex coordination between the two hemispheres of the brain. Earlier when discussing this aspect of development, we noted that there is a lack of knowledge about the period of two-handedness: One hand – then two hands – and still there is a unity. What a mystery! When we begin to discuss some observations of children who are perceptually disordered, the lack of knowledge becomes frightening. It is striking to note that *all* children who are perceptually disordered tend to grasp things with *only one hand*. They will use the right hand one time and the left hand another time – usually the one which is closest to the object.

S., 15 years and perceptually disordered, reaches for a piece of soap.

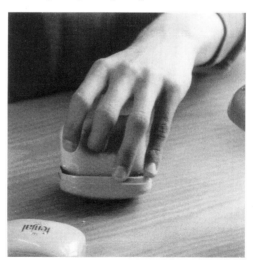

One-handed! He skillfully takes the soap out of the soap dish. But look how he does it. With the tip of his middle finger he lifts the soap. His index finger is extended in the air; his third and little finger touch the edge of the soap dish.

G., 10 years and perceptually disordered, washes the table.

He rinses the dishcloth in a bowl of water with one hand. His other hand rests passively on the table.

He wipes off the table with one hand while the finger tips of his other hand press against the table top.

Some time later, G. takes the crushed banana off the table with one hand while the other hand hangs passively at his side.

To open the milk container, he again uses only one hand.

It is surprising how often children who are perceptually disordered skillfully perform activities with just one hand. However, this is the impression we get when we watch only the hand and not the whole body. As soon as we observe the whole child, we become conscious of this *one-sidedness* – of using only one hand. Then we show concern by raising the questions: Why this behavior? *Why are children who are perceptually disordered functioning only on this elementary level of one-handedness? Why haven't they reached the level of using two hands at the same time?*

Unfortunately, using just the one hand is not the only troublesome aspect with regard to the movement patterns of touching as found in children who are perceptually disordered. When we observed G. wiping the table, we noticed that he touched the objects, but never really embraced them with his hand. What does it mean?

2.1.4 ...and Don't Succeed in Embracing Things

Let us again observe perceptually disordered children while they perform events of daily living. Notice especially *how* an object is held with the hand.

S., 15 years and perceptually disordered, washes a cloth.

◀

He grasps it with his thumb and middle finger...

and lifts it into the air with only two fingers of each hand: middle and ring finger of the right hand and the thumb and the middle finger of the left hand.

He lifts the cloth up to the height of his face - still holding it with two fingers. "Is it dirty"?

G., 10 years and perceptually disordered, always grasps objects with only two fingers.

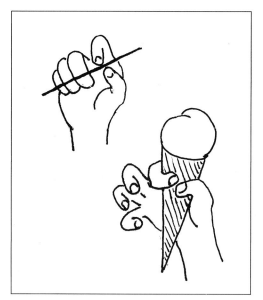

Here she uses only the thumb and forefinger; the other fingers are closed in a fist.
Now let's look at how she eats an ice cream cone. She holds it with just her forefinger and thumb; the other fingers are spread out. First she eats the cone from the bottom up to her two fingers. Then she eats the ice cream and cone down to her two fingers. Finally, she opens her grip and puts the rest of the cone into her mouth.

L., 6 years and perceptually disordered is removing the seeds from a grapefruit.

She touches the pulp with the forefingers of both hands; the other fingers are spread out.

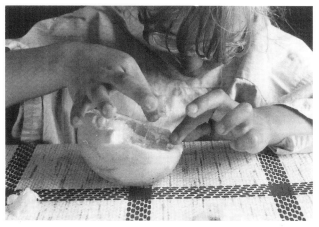

With the thumb and forefinger of her right hand, she removes a piece of the pulp while the other three fingers are spread out. The forefinger of the left hand presses the fruit toward the table.

With the forefinger and thumb of her right hand, she starts the grasping action – the position of a pair of tweezers.

With one finger of the left hand she holds the fruit, while one finger of the right hand searches for another seed.

Such observations support the conclusion that daily events always involve touching activities. Like other people, those who are perceptually disordered must touch and grasp objects in order to move them. During this kind of activity, they almost exclusively use only *two fingers* – the minimum number required to grasp an object. That's the least they have to do. What a difference from the grasps of those with a normal development. Instead of being able to embrace an object by using the whole hand, children who are perceptually disordered only take an object with two fingers. Like normal children, those with perceptual disorders also seek a change in resistance when grasping objects. However, normally developing children use all their fingers which they embrace around an object until they meet total resistance. In contrast, children who are perceptually disordered elicit a change in resistance with only two fingers holding the object in between. These two fingers seize the object like a pair of tweezers. They move toward it from both sides until total resistance is met.

Recall that in Part I of the book we described how children with normal development embrace objects on a support . Such experience is basic for getting to know the surroundings. When one can embrace objects in this manner, one can experience true three-dimensionality of things and receive important information about the qualities of persons and objects. Thus, only then can they become familiar with them. However, children who are perceptually disordered do not get this experience; they do not reach such familiarity with their surroundings.

A few examples showing this follow:

S., 15 years, is perceptually disordered. (See previous example p. 118.)

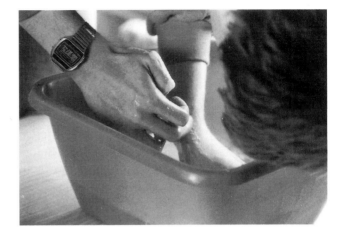

The cloth is in the water. He searches for resistance, does not find it, and becomes very tense.

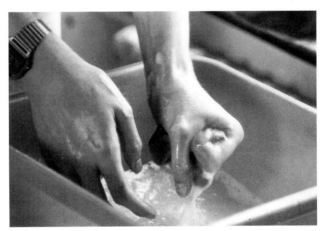

He seizes the cloth with one hand and presses his fingers together as much as he can. Where is the cloth?

C., 10 years and perceptually disordered, is removing seeds from an apple. To do this he has to bring the apple into the right position on the table. He seizes the apple and turns it.

At each turn, his fingers stretch out and are free in the air. His fingers are unable to follow the roundness of the apple. He seems to fail in correctly anticipating the movements of the apple and his fingers.

To conclude, the following sad example illustrates some consequences of being unfamiliar with the surroundings:

K., 9 years, has a severe perceptual disorder. She is in a home for mentally retarded children. I observe her eating a snack. In front of her on the table are a few cookies and a paper cup with juice. The other children around her are busy eating and drinking. K. starts with the first cookie. She grasps it and crushes it; there are small crumbs all over the table. The same thing happens with the second cookie. And the cup? She tries to drink from it, but it immediately becomes flat and the juice spills all over the table. Poor K.! During the time when other children around her are enjoying their cookies and juice, she has nothing. Finally, I approach her, even though I am only a visitor, and guide her hands so she can eat and drink.

The inability to hold and embrace objects results in a lack of experience with the qualities of the surroundings and, therefore, the persisting unfamiliarity with them.

To Summarize:

Children and adults who are perceptually disordered touch "the world". Their touching, however, causes them to *jerk away, become tense, and look away more often* than those who function normally. They know about the touching rules of the stable support, the sides, and the niche and do apply them.
However, when observing the movements of touching done by children who are perceptually disordered, it is striking to note that they use *only one hand* for many years longer than normally developing children. In addition, they more often utilize two fingers instead of five. They seize objects with the two fingers as if they were a pair of tweezers and therefore don't succeed in embracing the object. This is another sign that their experience about the world is lacking. As a result the world does not become a surrounding world for them (see Part IA, paragraph 1.2 "The World Becomes a Surrounding World: The World Embraces me; I Embrace The World").

2.2 They Know About the Rules of Acting Upon – But Where Is Their Changing of the Surroundings?

In Part I of the book, we described how children not only perceive the world but how they begin to explore acting upon the surroundings. To develop adequately, perceiving alone is insufficient. Nor is perceiving enough to solve problems of daily living. To do this, the knowledge about how one can act upon the surroundings is also necessary.
How do perceptually disordered children and adults perform in this regard? Do they know the rules of acting upon? Do they apply them? Where are the difficulties?
We will describe some observations of how perceptually disordered children and adults deal with the rules of taking off and of the neighborhood.

2.2.1 They Take Off – But How?

Do perceptually disordered children know about taking off? And if so, how do they apply that knowledge?

T., 8 years and perceptually disordered, visits his mother and newborn sister at the hospital. He sits on a chair and the baby is put on his lap. He is very quiet, holding her carefully and observing her attentively. He lifts one of her arms and lets it go. He becomes frightened when the baby's arm falls down and reacts by suddenly jerking his arms backward. Sometime later, T. holds the forearm of his sister and presses on it; he observes her little hands and opens her fingers. Then he looks into her face and tries to open her eyes. He repeats this performance several times during his visit.

D., 14 years and perceptually disordered is clearing a table. There are a few coasters on the table which have been wrapped in aluminum foil. He carefully removes the aluminum foil by scratching at it and puts it into the garbage can. Before he is finished, he also throws the coasters into the garbage.

These examples illustrate how children who are perceptually disordered explore their surroundings with regard to taking off what they see. Recall how we described normally developing children in Part I of this book. They explore their environment to make sure, for instance, that the nose of a person belongs to the face but glasses don't, or that colored spots they see as figures can be taken off the ground. In the above examples of perceptually disordered children, T.'s performance with his new sister and D.'s behavior with the coasters, are both quite similar to performances in normal children. The difference is that perceptually disordered exhibit this kind of exploratory behavior at a much older age than those with normal development.

It appears, therefore, that *children who are perceptually disordered do know about taking off.* They apply such rules in daily activities, even when it is difficult, as the following example illustrates:

L., 10 years and perceptually disordered, has taken the seeds out of a grapefruit (see previous example, pp. 119-120). When doing this, her fingers become sticky.

"Yuk! My fingers are so sticky"!

"How can I get rid of this"?

"This way"? It doesn't work.

"Or that way"?
That doesn't work either.

"Like this? No".

"Perhaps like this? Oh, this is so hard – and my fingers are so sticky"!

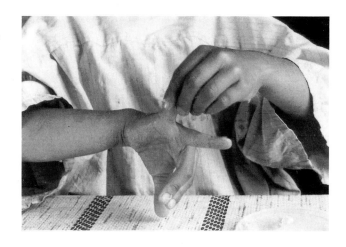

This example emphasizes how *difficult* it can be, even for an older perceptually disordered child to attempt to take something off. L.'s odd finger movements are a sign of the tension which is created when performing such an activity. They remind us of the finger movements made when such children touch something wet and cold (see paragraph 2.1.1 "They Withdraw From Touching, Become Tense and Look Away"). L.'s performance appears even more strange, though, because it involves not only touching, but also taking off.

The difficulties experienced by those with perceptual disorders become especially apparent when we reflect on normal children's performances. Their exploratory activity consists of the following steps:

- Exploring the relationships in resistance which may exist between a support and things one can see and feel at the same time.
- And in case that this resistance changes, trying to take that thing off a support and thereby changing the explored resistance.
- Attempting to release the thing one has taken off at another place on the support.

When children who are perceptually disordered attempt to apply the rule of taking off, their performances differ from children who are normal in regard to the number of different steps they are able to apply and how they are performed.

One group of children seems to explore only the relationship of resistance between the thing they can see and the support. Since many of these children are unable to grasp, they must use other movements to explore the relationship between the support and the spot they see. Some use knocking and/or hitting.

T., 14 years and perceptually disordered, knocks against objects with the nail of his middle finger, especially when he has nothing to do. As a toddler he did this for hours.

Knocking or hitting on the support can be observed frequently in hypotonic (floppy) children.

T., 9 years; M., 8 years; and G., 2 years, 6 months, were referred to us as children who are severely hypotonic. Whenever these three children are left to sit on the floor or at a table by themselves, they start to knock on objects which are within their reach.

A *second group* of children who are perceptually disordered tries to embrace and take an object off the support. Consequently, they perform one more step. But how?

K., 11 years and perceptually disordered, helps to scoop the seeds out of a tomato. I guide her. We cut off the top of the tomato. Then I guide her hands into the pulp so she can take out the seeds. As soon as she feels its mushiness, her fingers clutch the tomato and press on it so hard that it is smashed.

P., 15 years and perceptually disordered, stands next to his therapist who is wearing glasses. His parents are asking some questions and P. has to wait. Suddenly, he grabs the therapist's glasses and pulls them off his nose. P.'s fingers press around the glasses so strongly that the therapist fears they will break. Only with difficulty does the therapist succeed in loosening P.'s grip.

Many perceptually disordered children try to use their mouth to take something off.

C., 14 years and perceptually disordered, is helping to cook.
His fingers are sticky with dough. The teacher guides his fingers to his mouth so he can lick them. The middle finger touches his mouth first; he bites it. After several attempts he still can't manage to lick his fingers.

S., 16 years and perceptually disordered, is scraping the pulp from the skin of an orange. I guide his hands. We curve the skin a little and bring it to his mouth so the teeth can help. Grasping the orange skin, our fingers approach his mouth. As the pulp of the orange and our fingers touch his mouth at the same time, he bites down on my fingers with full force!

M., 5 years, 6 months, is perceptually disordered and hypotonic. For years he has tried to grasp whatever he sees, and if successful, to put it immediately into his mouth. During all this time, his therapist tried to inhibit him from using his mouth this way. She even advised the child's parents to do the same thing.

Knowing what we know, was this good advice?

T., 3 years, 6 months, is perceptually disordered, cerebral palsied, and severely hypotonic. I am playing with him; I guide his hands to squeeze a rattle which has the shape of a frog. We do this several times; then I let go of his hands. He grasps the frog and puts it into his mouth. Using his mouth he succeeds in squeezing and producing the rattling sound.

The use of the mouth in this way, seems inconspicuous enough, but for those who are perceptually disordered, it occurs more often and at an older age level than for normal children.

R., 10 years and perceptually disordered, wants to open a bottle of soda pop. He briefly tries to open it with his hands but without success.

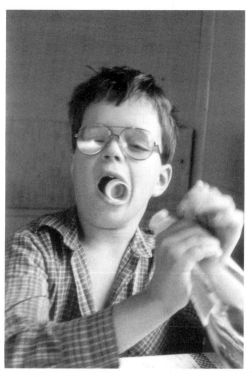

Now he puts the bottle in his mouth, bites the cap with his teeth, and pulls the bottle downwards. The cap is off!

A *third group* of perceptually disordered children tries to release the object they have taken off. However, their manner of releasing is quite different from that of normally developing children.

Ch., 4 years, 6 months, and perceptually disordered, has come for an evaluation. Several shelves in the room are filled with toys (educational materials), e.g., forms which can be inserted into each other, forms which can be piled up, and forms which can be matched. All of these toys have a great number of small pieces in different colors and shapes made of wood or plastic.
Ch. starts at the top shelf. He takes the first toy, a series of barrels, where the smaller one is inserted into the next one larger in size. He opens them up, removes the parts, grasps each one, and with an apparent elegance, throws them into the room, one piece after the other. He does this with every toy until the first shelf is empty. Then he starts with the second shelf, and for each piece, he uses the same elegant throwing motion.
At first his movements are quite controlled, but then his motion becomes faster and faster so that by the time he reaches the last shelf, his actions are quite frantic. In the end all the shelves are cleared, the floor is dotted with both small and large pieces of play material, and the colorful view reminds us of a meadow covered with easter eggs.

What has happened? Is Ch. simply a naughty boy who wants to get the attention of those who are around him? Is he in a late stage of defiance? Or does he need maximum changes in resistance when grasping and releasing?
In searching for an answer, we, too, can experience what happens when objects are handled with maximum changes in resistance. Try the following steps by yourself: Take a pencil or another object next to you and grasp it. Now press your fingers around that object as hard as you can - use maximum pressure with your fingers. Feel the maximum resistance from the object. Now, stretch your arm into the air. Release the object - but make sure that you do this with a *maximum change in resistance* in the *shortest time* possible. This means: Press hard; then change the maximum resistance into no resistance as fast as you can. Observe how you move your fingers. Isn't the movement of your fingers an elegant one? And what happens to the object? It doesn't simply fall to the floor. No, it "flies" through the room. Try these steps several times.

A *fourth group* of perceptually disordered persons perform not only the step of releasing, but in addition, the step of bringing together. The following examples illustrate this:

Mrs. H. who is brain damaged, is 36 years old. She has cut some apples and has put pie dough in several small baking pans. Now she begins to put apple pieces one after the other on the dough of the first pan. The other pans are next to the one she is filling. There are many pieces of apple within her reach on the table.
The first pan is filled, but Mrs. H. doesn't stop. Instead, she continues to take one piece of apple after the other and to pile them on the heap. The mountain of apple pieces grows; pieces begin to fall on the table. She takes them off the table once more and continues to put them on the top. She does this over and over again. Finally, she stops and looks desperately at what she has done.

What has happened? Obviously, merely looking at the heap of apple pieces does not help Mrs. H. to decide if there are enough, not enough, or too many pieces of apples in the pan.

C., 5 years and perceptually disordered, is making strawberry pastries. He uses small pre-baked, crunchy pie shells. The only action required is to fill them with the fresh strawberries. He takes one of the forms, puts in the first of the strawberries, and pushes hard on it - scrunch! The pie shell is crushed. With the next strawberry, he uses the same strong pressure and again the pie shell is crushed. C. continues like this. In the end he has crumbs with strawberries instead of strawberry pastries.*

What has happened? He was searching for maximum changes in resistance.

These examples illustrate that perceptually disordered children and adults know about taking off. Depending on the degree of their disorders, they perform one or more of the different steps which belong to taking off, and they search for changes in resistance.

What strikes us is the fact that children who are perceptually disordered begin to perform these steps at an older age than normal children, and obviously they, as well as the adults, attempt to elicit maximum changes in resistance.

2.2.2 And Where Is the Neighborhood?

The observations about recognition and production of neighboring relationships can be put into two groups: neighboring objects which touch each other and neighboring objects which do not touch each other and are only related over distance.

First, we will consider *neighboring relationships of objects which touch each other*. Perceptually disordered children can construct such relationships whenever *the situations are simple*, such as when a construction requires only *one kind of relationship* between *two objects*. The number of constructions of this kind occur so frequently that they are conspicuous, as in the following examples:

C., 5 years and perceptually disordered, discovers even the smallest opening in his environment, e.g., next to the radiator, in the wooden floor, or in the wall. As soon as he sees the gap, he tries to insert any kind of small object which he finds close by, such as, paper, cloth, buttons, small stones, wooden chips, etc.

K., 5 years and perceptually disordered, puts together as many pipes as he can find around. Then he connects his pipeline with a faucet and lets the water run through. Whenever he is by himself, he makes these constructions over and over again.

However, performances of perceptually disordered children disintegrate as soon as they require the handling of *more than one relationship* at a time.

D., 13 years and perceptually disordered, wants to prepare tea. He fills a tea ball with tea and closes it. He grasps it by the chain and hangs it on the teapot, but he puts it on the outside instead of the inside. He pours water into the pot but does not notice the ball is hanging outside. He looks into the pot and waits for the water to turn brown.

T., 10 years and perceptually disordered, is supposed to roll dough around sausages. He takes a piece of dough and one sausage and presses the sausage firmly on the dough. He prepares the next piece of dough, takes another sausage, puts it on the dough and presses hard. He continues like this until all the sausages are used but none of them are inside the dough.

F., 12 years and perceptually disordered, wants to prepare soup. The hot plate is on the table. The cord is wrapped around the hot plate. He pulls the cord by the plug and tries to put it into the socket. The cord is not long enough, and he doesn't notice that it would be long enough if he would unwind it. He pulls and pulls and finally gives up.

Let us review these examples. The child who is searching for gaps and the child who is connecting pipes are both dealing with two objects at one time. In the first example there is something that is open and something to put into it; in the second example, either two pipes or the pipeline and the faucet are involved – two objects in each case. The relationship consists of inserting something into something else. This action leads to a clear change in resistance, e. g., putting in until the movement is blocked totally.

We emphasize: in both of these examples the construction is a simple one. There is only one kind of relationship involving two objects. And in addition, in both examples the construction of such a relationship elicits a clear change in resistance. This may be the reason why perceptually disordered children can perform such actions.

Let's analyze the second set of examples. To prepare tea, D. needs to hang the tea ball *inside* the pot. This requires two relationships - the action producing the hanging of the tea ball *on* the pot, and at the *same time*, the action producing the hanging of the tea ball *in* the pot. D. cannot do this; he performs only one of the two relationships - the hanging of the tea ball on the pot. The performance with the sausages is similar. To roll the sausages *into* the dough requires two relationships: first, the action of putting the sausages *on* the dough, and then to place the dough *around* the sausages. This is too complex for T.; all he can do is press the sausages on the dough. F. should unwind the cord since it is wound *around* the hot plate. Now the relationship around consists of a series of neighboring relationships which have to be followed. Imagine winding the cord around the hot plate. One must first take the cord to the left side of the hot plate, then behind it, then to the right, again along the hot plate to the front, etc. The cord continually touches the hot plate at *another location*. This is too complex for F. - he only knows the cord and hot plate are touching each other. To get them away from each other, one can pull the cord and "being together"" will change into "being separate".

Common to all these examples is the *knowledge about the neighboring relationships of things*. Children with perceptual disorders usually can consider *one relationship at a time*, and even construct such a relationship, but they need *maximum changes in resistance*. Maximum changes are elicited by hanging the tea ball on the pot or pressing the sausages on the dough or pulling the cord off the hot plate. Depending on the particular problem, this construction of events allows for either solving or not solving a problem. The first two examples showed success; the next three examples showed failure.

Neighboring relationships also become *more complex* when they include *more than two objects*, as in the following examples:

O., 14 years and perceptually disordered, is supposed to hang the wet clothes on the clothesline. He is guided at the beginning of the event. A towel is hung over the line, then attached to the clothesline with two clothespins. For the next piece of wash, O. takes a clothespin by himself and holds it against the piece of cloth where it touches the line. Of course it does not hold. O. notices this. Now he tries to put the clothespin upside down over the line. The clothespin falls to the ground. For his third try, O. attaches the clothespin the correct way.

A., 11 years and perceptually disordered, is going to clean a room with the vacuum cleaner. She puts the parts together. She looks around with the cord in her hand. Finally, she inserts the plug into a key hole.

She goes to the butcher store to buy meat. As is usual in Switzerland, she takes a shopping bag with her. The butcher gives her two packages. She puts them both under her arm. Soon, the packages begin to slip and almost fall to the ground. A. has to keep on adjusting them so they don't slip out. She doesn't realize that she could put them into her shopping bag.

These examples include neighboring relationships involving *three* objects: O. hangs the clothes on the line and tries to fasten them with clothespins. To do this he has to establish relationships which include the clothespins, the clothes, and the line. At first he *touches* the piece of clothing with the clothespin. This causes a relationship between two objects with a change in resistance. Later he tries to put the clothespin on the piece of clothing. This causes another relationship between two objects with a change in resistance. Only on the third trial does he succeed so that the clothespin fastens the piece of clothing and the clothesline at the same time. This finally causes the relationship which includes three objects. A. searches for the socket to plug in the cord. While doing this, she considers only two parts of the problem – the cord and an opening. Because of this, she is unable to solve the problem. Solving this problem requires an additional, third part – that of having the electric current which is available only at an electrical outlet, not a key hole. A. purchases meat. She holds the two packages under her arm; she establishes a relationship between her body and each separate package. This corresponds to the criteria for a simple neighboring relationship. She fails in the more complex neighboring relationship involving herself, the shopping bag, and the two packages *at the same time*. This neighboring relationship would require that she put both packages into the bag and carry the bag home.

Now, let's consider the *neighboring relationships* of objects that *do not touch* each other *directly*, but share a *common support*. Touching is only an indirect possibility. How do children who are perceptually disordered perform under such conditions?

To answer this question a second group of observations will be used.

M., 16 years and perceptually disordered, prepares a special kind of cookie. Each one consists of a marmelade filling between two round layers of dough.

The baking sheet is nearly covered with cookies; there are only a few spaces left.

M. cuts out a round piece of dough and tries to put it on the sheet. Instead of being cautious as he puts it down, he lets it fall on the sheet. It falls partly on another piece of dough and partly on the open space. He is able to slide it into the gap, but for the next pieces of dough, he repeats the same performance. In spite of being told to be careful, he isn't. He continues to drop the pieces anywhere on the sheet instead of sliding them into the open spaces.

MR., 11 years and perceptually disordered, is having a rhythm lesson. He is supposed to step over straw mats on the floor without touching them. He cannot perform this action. He steps on the mats instead of in the spaces.

B., 14 years, is perceptually disordered. His mother reports that he can recognize the smallest objects from a great distance. For instance, he loves babies and becomes all excited when he sees a baby carriage, even when it is still so far away that it can hardly be seen by anyone else.

S., 3 years and perceptually disordered, sees a basket of nuts on a round table. He had played with them a short time earlier. He tries to grasp the basket but it is too far away. He begins to walk around the table and stops from time to time, trying to grasp it, but it is still too far away. He goes around the table in this manner several times and tries over and over again to grasp the basket. It appears that he is unable to judge that the basket is too far away.

In the first two examples objects are in an indirect neighboring relationship over a support. The third and fourth examples include a relationship between an object and the body, where the object can be touched only indirectly over a common support. In all four examples it appears that children who are perceptually disordered have difficulty in properly evaluating neighboring relationships. M. doesn't find the spaces between the cookies; MR. steps on the mats instead of the spaces; B. cannot evaluate how far away the baby carriage is; and similarly, S. does not seem to realize that the basket of nuts is out of reach.

The difficulty in *evaluating distances* between objects and between our own body and objects that are not touching each other directly, is also expressed in the following examples:

J., 5 years and perceptually disordered, plays outside the house. At the entrance there is a door sill. From time to time he tries to enter the house, but each time he stops his movements and prepares to step over the sill when he is still too far away.

D., 12 years and perceptually disordered, has a very hard time judging distances. When he goes shopping he has to cross a church square. There are several stone pillars at hip height which surround the square. Most of the time D. runs toward the pillars until he almost bumps into them, stopping only at the last second. One day, when he is having a hard time, he runs toward the pillars which such force that he bumps right into one and the force throws his upper body right over it.

Another time he is on a giant Ferris wheel at the fair and appears not to be aware of heights. Riding with his teacher and a classmate, their passenger car reaches the highest point and stops. The classmate holds on to the bars for dear life and anxiously peers over the edge. D., however, looks down without any fear and shouts, "Look, you can see the Schutzengarten (a very large building)! We pass by there on the way to school".

AW., 9 years and perceptually disordered, is hiking with us at camp. We come to an electric fence. I help AW. under it. He shouts, "Ouch," even though he is not touching the fence.

D., 10 years and perceptually disordered, takes part in a game of mastering obstacles. He is supposed to pass underneath a rope that is stretched one meter high across a path. D. only needs to bend a little bit to do it. Instead, he gets down on all fours and crawls. It is obvious that he cannot judge the distance between the rope and the ground.

C., 12 years and perceptually disordered, helps to prepare jam. The fruits are soft and need to be pressed through a sieve. With his fingers he pushes the pulp of the fruit through the sieve. After a while he turns the sieve over so he can lick off some of the pulp. As he does this, some of the pulp falls on the table. He doesn't appear to notice. More fruit is put into the sieve and again pressed and pushed through. C. is attentive while doing this. After a while he turns the sieve to again lick off some of the pulp. Once more some of the pulp falls on the table. This time he notices it and seems to be astonished that there is pulp on the table. He doesn't appear to connect the pulp on the table with the pulp in the sieve or to connect it with what he is doing.

In the last example the child did not appear to establish a neighboring relationship between the fruit pulp in the sieve and that on the table.

Acting upon happens in *time*: one activity follows the other; the effect appears after a causative action. A sequence or *series* of events becomes important. How do children and adults who are perceptually disordered deal with sequences of events?

2.2.3 The Sequence - When Something Is Missing or When You Cannot Go Back

The first observations describe the difficulty in following a certain sequence of activities. We often notice that children and adults who are perceptually disordered *omit certain steps.*

R., 10 years and perceptually disordered, is in therapy. His therapist wants him to squeeze a lemon. In preparation for this, they put a small lemon juicer together by pressing the upper part tightly on the lower part. Eagerly, R. looks at the lemon juicer as they finish putting it together. He asks, "Why isn't there any juice"? The lemon is still on the table next to the lemon juicer.

D., 14 years and perceptually disordered, takes some peas out of the pods. He then puts them into a pan. Afterward he gets a hot plate, inserts the plug into the socket, puts the pan on the plate, and turns the heat on. After a while the peas begin to burn because he has forgotten to add water.

After swimming, T., 11 years and perceptually disordered, sits under the hair dryer with a swim cap still on his head. Only after some time passes does he take off the cap.

R., 10 years and perceptually disordered, tries to make confetti with a paper punch. He puts the paper into the punch and presses over and over again but forgets to slide the paper to different positions. After a while he is surprised that there is no confetti.

A second group of examples illustrates another difficulty often observed. Somewhere in a sequence an *error* is made. To solve the problem, the error needs to be *reversed*, but how?

K., 6 years and perceptually disordered, is busy placing disks on a stick by putting them in order in decreasing size. She makes an error but does not notice it until after having put another disk on the top of the one incorrectly placed. She sits helplessly in her chair. Finally, she turns the stick upside down so that all the disks fall on the table. Then she starts all over again.

The behavior of *starting all over again* is frequently observed. It is a sign of rigid representation of sequences. To begin again, however, is not always so easy and can create its own difficulty, as shown in the next example:

C., 10 years and perceptually disordered, is attaching a wooden board on the wall as a shelf. It will be next to a window. His therapist is guiding him. They drill two holes in each end of the wooden board; then they put a cord through the holes and attach the shelf by the cord to the wall. Having finished, they notice the window can no longer be opened because the shelf is too large.
C. does not know how to correct this. The therapist proposes cutting off part of the shelf. C. disagrees and insists that by doing this, he will lose the two holes and will no longer be able to hang up the shelf.
Reluctantly, though, he allows the therapist to guide him as follows: Remove the shelf from the wall; get the saw and saw a part of the board off. As soon as the piece of wood is sawed off, he declares, "Ah! Two holes can be drilled and a cord put through". Quickly he starts to perform those steps.

It appears that C. can only imagine that the two holes will be gone when a part of the board is removed. This means he can represent the next step, but he cannot imagine the steps which come after that, such as the drilling of two more holes. In other words, he cannot represent the whole sequence of steps involved by starting again. Only when he has reached the point where the actual situation is present, i.e., the newly cut off board without holes is right in front of him, can he know that it is possible to drill two new holes.

To represent the order of activities in a sequence and to carry them out is very complex. For instance, when I am carrying something with both hands and need to open a door, I have to put down the objects I am carrying with one hand so that my hand becomes free. Now I can open the door with the free hand, pick the object up off the ground, walk through the doorway, put the object down again, and shut the door. Afterward I can pick up the object and proceed on my walk. To perform such ordering of activities in time can be very difficult or even impossible for children and adults who are perceptually disordered.

B., 9 years and perceptually disordered, is trying to paste pictures on a piece of paper. She takes a stick of glue and opens it. Then she sits there with the sticker in one hand and the *glue in the other. How shall she proceed? What comes first? What next? And then what? B. becomes desperate – she panics.*

Mr. R. has suffered a head trauma. He is in therapy and is trying to prepare dough with an electric mixer. He is holding the mixer in one hand. He knows that the electric cord has to be connected to an extension cord. This demands difficult manipulations with the use of both hands at the same time. The sequence of steps would be as follows: First put the mixer down so that one hand becomes free. Then hold the extension cord with one hand so the other hand can insert the plug into the socket. Next, let go of the extension cord; take the mixer off the table. Finally, start the mixer.
But Mr. R. just sits there and looks at the mixer he is holding in one hand and at the plug he is holding with the other. He does not know what to do next.

In clinics where there are adults with severe brain damage, I often hear complaints that these patients do not call the nurse during the night when they need to go to the bathroom even when they are capable of ringing the bell. To explain this kind of failure, I point out that they have difficulty in sequencing daily life activities. In situations like this, after first feeling the need to go to the bathroom, the patient must represent the following series: When I ring the bell, the nurse will hear it; when he or she comes to my room, I will be helped to get up and go to the bathroom.

2.2.4 When Only the Moment Exists...

The activities of children and adults who are perceptually disordered are often related to the immediate situation and *do not include what may follow.*

M., 15 years, is perceptually disordered. When he is building something, he closes the box of nails each time he removes a nail, even if he is going to need several of them. Similarly, he takes the drill out of its case to use it and then puts it back, even if he will soon need it again. He closes the bottle of juice after pouring some into his glass, even if he sees his neighbor's empty glass and his neighbor has not yet received any.

This difficulty in representing what will happen afterward causes another dilemma: that of *not being able to wait.* This problem is difficult to understand on the part of family members or other people.

B., 5 years and perceptually disordered, has taken the train many times and has never been late in catching it. Also, he has quite often accompanied visitors who are departing from the train station. The trains originate from this station, and so this means that they sometimes have to stand around for quite some time before departure. In spite of this, scenes like the following often occur:
B. accompanies his father and me to the station. We arrive early. When we enter the train, B.'s father puts the luggage on the rack and we begin to talk. Soon B. becomes restless. "The train will be leaving," he says. His father explains to him that there is still much time left because the conductor has to check the brakes on each car, the mail has to be loaded, etc. His father suggests that he go and watch all of these activities. "Also," he tells him, "the stationmaster is still standing in the station office; he has not even put his cap on his head yet". He reminds B. that the stationmaster always puts his cap on first and then steps outside onto the platform to look around and make sure everything is all right. In the end the conductor will close all the doors and shout, "All aboard"!
His father tries to explain to him how all of these activities will occur before the train leaves, but it doesn't help, as B. becomes more restless. He declares that he will go out of the car. No one reacts and so he begins to shout, "Out, out"! Still there is no reaction from the adults. Then he begins to cry and stamp his feet.

Is B. a spoiled child? How can his behavior be explained? Why does he have this difficulty of not being able to wait? Why is the representation of happenings in his environment so poor? Children with perceptual disorders are often described as being poorly oriented in time and space.

E., 8 years, is perceptually disordered. When someone explains to her in the morning that she will go skating or swimming in the afternoon, she goes to get her skates or swimming suit and wants to leave right away.
At other times, for instance, when visiting friends with her mother, they were invited for dinner. After dinner, the adults sat around the table and talked. Her mother turned to her and said, "You can play with your ponies and your doll. When it is dark, you can put them into your bag, and we will go home". She immediately put her ponies and doll into her bag. She got her coat and was ready to go home right then.

Another difficulty is related to the preceding ones. *Those who are perceptually disordered are often afraid of the dark.* Parents report that they have to leave the lights on during the night or their child would never go into a dark place, such as the basement. They tell me that their children have a very strong imagination, seeing ghosts everywhere. Similar behavior can be observed in children who are normal; the difference is in the extent or magnitude of that fear. Children who are normal can orient themselves surprisingly well in a room, such as, their bedroom, even in the dark. They can find a handkerchief under their pillow exactly where they had put it two hours earlier. Children who are perceptually disordered will show specific difficulties in this regard as illustrated in the next example.

D., 14 years, is perceptually disordered. A teacher and D. are carrying a case of bottled mineral water to the basement. The rooms downstairs are semi-dark. The teacher guides D.'s hands to switch on the light. D. begins to talk excessively, as he always does when he is afraid. He comments on what he is doing with expressions like, "Find the light. There is something. Ouch! Almost fell down. Ouch! So dark. Found it. Now the light is on".

Since only the present moment exists, perceptually disordered children have difficulty assessing and detecting danger.
Where does danger lurk? Even though children who are perceptually disordered are afraid of dark corners, they do not usually detect true danger and may go in any direction or hide in the farthest corner of the room when the slightest movement occurs.
When out for a walk, some children who are perceptually disordered continue walking without turning around to watch where the others have gone. Some do not even react to calls. It is a big developmental step for them when they finally learn to stop when called. There comes a time, too, when these children begin to see a likely danger from afar and react as if it were right there. For example, they may be frightened of a dog, even though the dog is still much too far away to do them any harm. Can we suppose that they are unable to judge the distance? (See Part II, paragraph 2.2.2 "And Where Is the Neighborhood"?.)

T., who is perceptually disordered, has grown up on a farm. As a toddler, he would run through the middle of a cow herd, apparently without noticing the animals; he would crawl into the cribs, even when the animals were eating out of them. Once he was bitten by a dog, but this did not change his behavior.
When he was 11 or 12 years old, his behavior changed. He began to be tense when he approached an animal and his face would become filled with anxiety.
At the age of 12 years, 6 months, he was again bitten by a dog. From then on he reacted with great panic to animals, e.g., even when a dog was at a great distance or when he only heard one bark. After once seeing a dog near a particular house, he was afraid to pass by that house.

From a great distance T. notices the house with the dog; he refuses to continue his walk.

When he reached the age of 13 years, he learned to adapt his behavior to such situations. Now when he encounters an animal, he climbs on a fence or takes the arm of an adult. At home when the cat has left the room, he closes the door so it cannot come back in.

2.2.5 ...and Causative Actions Do Not Correspond to the Situations

The solving of daily problems requires the use of causative activities in order to elicit certain effects. Several disturbances may affect such activities. One may use the right action to cause an effect, but the manipulation may not be adapted to the actual situation, and therefore the effect is not reached. That is, the changes needed to solve the problem do not happen. Most of the time children who are normal notice when this occurs; they change the manipulation and examine the effect. One can also get an undesired effect because of using the wrong action. Then the questions arise: Is the error noticed? Can it be corrected?

D., 14 years and perceptually disordered, tries to close the door of a cabinet. There is a hook on the cabinet which can be put into a specific position so that it blocks the door from moving any further.

D. turns the hook several times, and at the same time, presses his knee against the door so it doesn't open. He continues turning the hook, but when he removes his knee, the door is still open. He starts turning it again; he doesn't know when to stop.

Obviously, D. cannot judge the effect of his turning the hook. Perhaps he misses the change in resistance?

C., 12 years and perceptually disordered, wants to lock the door of the schoolroom during lunch break so nobody can enter the room.
While he is alone in the room, he ties a string on the doorknob. Then he fastens the string to a desk located next to the door, but since the door opens toward the inside of the room, the desired effect is not elicited when another student enters the room. After that student leaves, he pulls a big box near the door and fastens the string to the box. He still doesn't get the desired effect. Now he begins to pile

chairs and boxes against the door until they reach quite high. Some time later the teacher enters the room, and they all tumble down with a loud crash. C. laughs.

It is raining. The class is going on a field trip. T., 14 years and perceptually disordered, puts on his rain coat and pulls the hood over his head. At the door there are a few umbrellas, and T. grabs one. The teacher puts the umbrella back and says, "You have your hood on". T. takes the umbrella again, opens it, and holds it over his head during the whole trip, even though his hood is still on his head.

K., 11 years and perceptually disordered, wishes to hang up her picture. She takes a hammer and a nail, goes to a window instead of the wall and tries to put the nail into the glass.

A., 11 years and perceptually disordered, is making vanilla pudding. To do this she needs to open a carton of milk. She cuts into the middle of the carton with a knife and milk spills out. She appears to be surprised.

Each of these children applied a *correct causative action* – D. turned the hook, C. took a string, T. grabbed an umbrella, K. tried to hang up a picture with a hammer and nail, and A. tried to open the milk carton with a knife – but *none* of them *adapted their actions to the specific conditions of the given situation.* D. did not find the correct position for the hook; C. failed to represent the neighborhood relationship between the objects in the room and the movement of the door; T. did not notice that his hood already protected him from the rain; K. did not realize that it was impossible to hammer a nail into a window pane; and A. did not expect the milk in the box to squirt out from the pressure of the opening.

The next set of examples illustrate how incorrect effects are elicited by children who are perceptually disordered because they apply *the wrong actions* to the situation.

M., 16 years and perceptually disordered, takes the train to school once a week. He has a commuter card which has to be punched each time he makes the trip. One week he has no trip to school and declares, "Next time I must punch it twice".

T., 14 years and perceptually disordered, is preparing lunch. The therapist who is guiding him helps him turn the stove to the highest temperature in order to heat the instant soup. The therapist has a few things to do and doesn't pay attention to T. for awhile. Suddenly the therapist smells food burning and runs back to the kitchen where she finds T. standing at the stove, vigorously stirring the soup. He is waving his other arm nervously and his facial expression shows panic as he utters, "Aaugh! Aaugh! Aaugh"!
Obviously, T. had not noticed that the reason for the problem was the overheated burner and not the manner in which he was stirring the food.

When a desired effect doesn't happen, children who are perceptually disordered often try to change the causative action. However, the inability to adapt their manipulations may cause another undesired effect. When they search for a better solution, they give the impression that they are very busy, as is illustrated by the next example:

S., 5 years, 6 months and perceptually disordered, boils an egg for sandwiches. As the water begins to boil, steam comes out from under the lid. She wants to remove the lid to take the egg out. She grabs the knob on top of the lid but burns her hand.
Next, she tries to make the lid slide off by using a wooden spoon. However, she holds the handle of the spoon close to the bowl, and when she moves it toward the pan, she again burns her hand. Now she takes a small plastic bag, puts it over her hand and grabs the lid. Again she burns herself – the bag melts.
She looks around and appears to be thinking. Then she gets an apron from a hook. On her way back to the stove, she wraps the apron

around the handle of the spoon and holds it with her right hand. With her left hand she tries to make the lid slide off the pan and burns her hand another time.
She shakes her head. She now wraps the apron around her right arm and hand, grabs the spoon and pushes the lid off the pan with that hand. She can now breathe a sigh of relief.

S. gives the impression that she is very busy – very active. She experiences the hot steam and its effect; she gets burned. She has to avoid that effect. But how? She takes a wooden spoon, but holds it in the wrong place. Somehow she seems unable to anticipate how to manipulate it. This also appears to be the case when using the plastic bag. She doesn't recognize that it will not give protection or that it will melt. Afterward she wraps an apron around the wooden spoon. Getting the apron and wrapping it around her arm in order to be protected from the heat could be a correct causative action, but the situation was too complex. Instead of wrapping her arm, she wraps up the object she had just used, the spoon. Then comes the next obstacle: She uses her unprotected hand to remove the lid. We can infer that S. cannot consider the network of relationships involved in the actual situation. Thus, she makes short circuits. The lid has to be moved; a spoon might help. The steam is hot; an apron helps to protect from being burned. Therefore, she puts the apron and spoon together. Only after being burned again is she able to change her performance and finally reach her goal.

2.2.6 ...Then the Wirklichkeit Slips Away

Daily life is continually changing. Even so, we sometimes get the impression that our life is monotonous. How is the behavior of those who are perceptually disordered affected when such changes occur in their daily lives?
One peculiarity is that they get *excited* very quickly.

D., 11 years and perceptually disordered, often takes the trolley with his classmates. Usually they have to get off at Market Square, but today they must go two stops further. When they reach Market Square, D. becomes panicky and shouts, "But we have to get out here"! He hits his face and begins to appear desperate.

During the first few sessions of therapy, D. and his therapist sit in the same place every time. Then one day the therapist changes her place. D. begins to scold, "You should be sitting here, not there; I want to sit there".

Another peculiarity is the tendency to *reconstruct the surroundings to what was seen before a change occurred* (see also Part I B, paragraph 1.1 "I Restore the Wirklichkeit").

T., 11 years and perceptually disordered, is with his class on a walk. They pass by a fence made of iron bars. One of the bars is rusty and broken. T. lets go of his teacher's hand, lifts up the broken part, and holds it against the part which is still intact.

Some children take this need to have their surroundings not change to an extreme. They almost become compulsive. For some, doors must always be shut; the zipper on a jacket must always be closed; and even if their sleeves get dirty or wet when working, the sleeves must never be pushed up.
Perceptually disordered children don't want changes to ever occur in their familiar surroundings, and they want events to always happen the same way. This highly conservative behavior is also expressed in the activities which they can perform themselves. Once they have grasped the order of an event, they will always follow the same procedure. Therefore, these activities will appear to be *monotonous,* and their behavior, meticulous.

A., 8 years and perceptually disordered, is the first child to arrive at school in the morning. As soon as she enters the building, she always performs the same sequence of actions: She turns on all the lights and closes all the bathroom doors with a loud bang. If someone tries to interfere with her actions, she begins to shout and stamp her feet.

B., 9 years, is perceptually disordered. During lunch break, the teacher puts photographs from a camping experience into an album with self-adhesive photo corners. The foil backing must be removed and thrown away later on.
These little pieces of foil are now all over the table. B. begins to take them off, and as he works, he becomes more and more meticulous about cleaning the table. He discovers other things – points of pencil lead, rubber bands, a pencil, and scissors. He wipes away the pencil points, puts the rubber bands back, and shoves the pencil and scissors to the other end of the table.

When a *situation changes, the performance deteriorates rapidly*. Often the practice of specific actions with children and adults who are perceptually disordered, may allow them to reach a certain level of skill. However, once outside the classroom or therapy room, performance deteriorates. Then we hear complaints about poor transfer of the performance.

R., 35 years, has a head trauma after having had an accident. In physiotherapy she has learned to walk again. This has taken a long time and a lot of practice. She can walk, and yet, she cannot walk, for often her performance breaks down as soon as there is the smallest change in the situation.

In her case, a particular movement was always practiced in a specific situation – in the therapy room. There are very few changes in a therapy room. This is the difference between therapy and daily life situations where there are unexpected noises; uneven ground; furniture and other obstacles in the way; and people present who move around, call, or want something.

M., 9 years and perceptually disordered, is with his class in a gymnasium. Two perpendicular bars are set up, and a rope is attached between them so the children can jump over it. M. knows how to jump, but as soon as he is close to the rope, he becomes all excited, makes facial contortions, and refuses to jump.

T., 9 years and perceptually disordered, begins to construct a wooden grate on the table. Before it is finished, it is time for lunch, and so T. is supposed to clear his work off the table. He becomes angry. He picks up the grate he has started and breaks it apart.

For both children, the situations changed. M. knows how to jump in a specific situation, but with the construction of the bars and the rope in between, the situation is an unfamiliar one, and he can not jump. In the case of T., a new situation also came up. He was making something and using a table for the work surface. Now, the table with all of his work on it has to be cleared. He cannot accommodate this new condition and becomes tense.
Tenseness also occurs rapidly in situations when an *action does not immediately lead to success.*

C., 12 years and perceptually disordered, wants to squeeze an orange with his hands. He cuts it in half and puts one part into his mouth. He doesn't succeed in sucking the juice out of it. His face becomes distorted and he shouts, "Can't do it"! He throws the orange away.

S., 13 years and perceptually disordered, is sawing a wooden board for a cupboard shelf he is repairing with his teacher. The sawblade breaks. S. becomes tense; his face becomes distorted. The blade must be changed. He tries to position the key that loosens the screw but doesn't succeed. He runs off and begins to hit himself. The teacher guides him through the manipulations and he quiets down.
The new shelf is ready to be put into the cupboard. S. has fixed holders at three ends of

the shelf. (He has overlooked the fourth one.) He puts the shelf into the cupboard. It holds for a moment, and so he begins to return some of the things that belong on the shelf. The board starts to tip. He becomes tense again and throws the two boxes of nails he was putting into the cupboard onto the floor.

Repeatedly, when situations change performances of perceptually disordered children and adults deteriorate. Each time, it is as if *the Wirklichkeit is slipping away*. This conflicting behavior and these tensions, which are created when performances break down, are very difficult for family members and others around them to understand. They ask, "Since they have been able to do it before, why is it that they suddenly can't do it"?

To Summarize:

Children and adults who are perceptually disordered *do touch*, but *not in the same manner* as those who are normal. They jerk away and cannot look at what they are touching. Often they grasp with only two fingers and use only one hand. At other times though, they use both hands.

They *act upon* things, and yet they *do not act* like normal children and adults. They take things off, but they leave out certain steps. They press until the material breaks, or they throw things and hit on them. They establish neighboring relationships when the objects touch each other but cannot seem to do so when they have to reach for them across a support. They can use visual information quite well but cannot judge distances.

They have a *problem with sequences*. Why are sequences so difficult for those who are perceptually disordered?

How can we explain these peculiarities? Does the deterioration of performances have anything to do with motivation? Or, is the explanation not so simple? We will try to answer these questions in the next chapter.

3 What Happens When There Is a Lack of Tactile-Kinesthetic Information?

The observations of Part I, "Living in a Wirklichkeit," illustrated how normal children become familiar with their environment and acquire knowledge about the Wirklichkeit. During this period they are constantly moving. They move their eyes to gain visual information. They turn their heads toward noise sources to receive auditory information. They change the positions of the body – the trunk, the head, the hands, the fingers, the mouth, the arms, and the legs – in order to perceive tactile-kinesthetic information.

It is in this manner, that normal children continually search for information.

3.1 They Search for Information

We need information in order to develop and learn. We have described the behavior of perceptually disordered children and adults in this part of "Failing in a Wirklichkeit". It becomes important, therefore, to analyze how their behavior reflects their *search for information*.

Furthermore, it is important to consider what *kind of information* is being searched for. Information can be visual, auditory, and tactile-kinesthetic. What kind of information is the perceptually disordered child or adult picking up?

3.1.1 They See and Hear

Many activities of children who are perceptually disordered are oriented toward eliciting visual or auditory effects, as is shown in the following examples:

Activities with Visual *Effects*

R., 10 years and perceptually disordered, is drilling holes with a hand drill. He turns the handle wildly, watching how it whirls. To us he calls out, "Look Mommy! Look Patrick (his classmate)"!

C., 9 years and perceptually disordered, is in a storeroom where a lamp hangs down on a long cord. He is holding the lamp-shade with one hand and asks, "Can I"? Before anyone can answer, he quickly twists the shade around on its axis like a merry-go-round, lets it go, and enjoys watching it twirl.

Activities with Auditory *Effects*

R., 10 years and perceptually disordered, wants to get a hammer which is hanging from a hook on the wall. He touches the hammer, intending to take it off, but it slides out of his fingers and bangs against the wall instead. This produces a loud sound. R. repeats the movement several more times, obviously trying to elicit the sound again. He even forgets that he had originally wanted to get the hammer.

He takes a small ball and lets it roll down a channel he built for himself. However, the ball rolls off. He gets it again. Now he attaches a can at the end of the channel so the ball rolls into it. He seems to be fascinated by the sound created when the ball falls into the can because he begins to vary the bounces of the ball so that there are different vibrations.

Normal children also produce activities which cause visual or auditory effects. The effect may be a new one, and the children will repeat the movement to elicit the new signal again. Soon their curiosity is appeased, though, and they go to another activity. Such behavior is mostly observable during the first year of life.

All children, therefore, search for visual and auditory information, but the constant repetition of such activities is peculiar to children who are perceptually disordered. Also startling is the age level at which they perform such activities – much older than normal children.

This pronounced orientation toward visual and auditory information is expressed in several of the observations that were described. In paragraph 2.2.6 "...Then the Wirklichkeit Slips Away," we described the panic of D. when his classmates would not leave the trolley at the usual stop but went two stops farther. In daily life, we decide when to get off the trolley based on specific visual information. D. does this, too, but the unusual characteristic about him is that he panics when there is an exception.

A similar analysis can be made of other observations.

During several sessions, D. had observed that the therapist always sat on a specific chair at a specific location in the room. He stored the situation as if it were a photograph. He could not endure that this visually stored situation changed. T. holds the broken part of an iron bar of the fence up to the other part which is still in place. With that gesture, he reproduces a previously stored visual picture of the fence. In the same paragraph 2.2.6, we mentioned how children who are perceptually disordered meticulously clear tables. We can also interpret their activity as an attempt to restore visual impressions of the surroundings as soon as they are changed. The table is usually a large plain surface. Pieces of foil, points of pencil lead, rubber bands, etc., are colored spots on that surface. Colored spots, however, disturb the familiar visual picture of the monochromatic table. Therefore, they have to be wiped off.

This kind of behavior differs from that of normal children. Piaget and Inhelder (1956) investigated the development of representation, e.g., spatial concepts, in normally developing children. Their findings contradict the model of a photograph. This means that normal children select visual information in a given situation, but they relate the visual information to tactile-kinesthetic information which has been experienced when interacting with the Wirklichkeit.

3.1.2 They Receive Tactile-Kinesthetic Information

We described how children and adults who are perceptually disordered apply the rules of touching and of acting upon. They *exert much force* in their activities, thereby creating the impression that they are aggressive. Those around them in the environment are not aware that they are searching for *information* by eliciting maximum changes in resistances.

We presented examples of *touching*. In paragraph 2.1.2 "They Know About the Rules of the Stable Support and the Side," we described P., who sways back and forth when standing. In this way he elicits, over and over again, maximum changes in resistance between his feet and the support. This search for information lets him know where his body and the support are. C. searches for the same kind of information when he presses his hips against the edge of the table while he works. S., too, elicits a maximum change in resistance when he presses the wet cloth into his hand with great force. Also, K. in the home for those who are mentally retarded looks for maximum changes in resistance when she crushes the paper cup while drinking and crumbles the cookie while eating (see paragraph 2.1.4 "...and Don't Succeed in Embracing Things).

We presented examples of *acting upon* in the section "They Take off - But How"? (see paragraph 2.2.1). We inferred that children who cannot grasp, e. g., those who are hypotonic or those who knock and hit on a support try to change the resistance between the support and what they see on the support. Perceptually disordered children who can grasp try to separate different colored spots they see using more than normal strength. P. grasps his therapist's eyeglasses with such force that the therapist is afraid they will break. K. wants to take the seeds out of the tomato, but instead, her fingers smash the fruit in the attempt.

Many children who are perceptually disordered use their mouths in order to elicit maximum changes in resistance. They bite what they feel with extreme pressure. C. tries to lick the dough off her fingers and, in doing so, bites on them. S. uses his mouth to scrape the pulp from the skin of an orange and ends up biting his finger. Often, children who are perceptually disordered have difficulty licking things since there are no maximum changes in resistance for that action. Instead of licking an ice cream cone, they bite on it. They also need maximum changes in resistance when eating. Therefore, they fill their mouths until there is no space left (see paragraph 1.4 "They Are Labeled Ill-Mannered").

Children and adults who are perceptually disordered attempt to elicit *maximum changes in resistance in a multitude of situations.* Ch. grabbed toys and threw them through the air in an elegant way. C. pressed the strawberries so hard on the crisp surface of the dough that the pastry crumbled. (See paragraph 2.2.1 "They Take off – But How"?) D. wanted to communicate something to his classmate in the pool where they were swimming, but he could not call her because it was too noisy. Therefore, he wanted to touch her to get her attention, but how does he know when he is actually touching her? Only when he has elicited a maximum change in resistance. The panic is there. The search for information begins. A similar situation is the one with the car antenna. D. wanted to know if the antenna belonged to the car or not. Therefore, he touched every antenna he saw. How did he know when he touched the antenna? Only when he elicited maximum changes in resistance. By then, the antenna was broken (see paragraph 1.3 "They Are Called Aggressive").

There are many situations in which one cannot elicit maximum changes in resistance. This causes a lack of information and results in deterioration of the performance. The desired goal is not reached, and a problem of daily life remains unsolved. As Mrs. H. piled up pieces of apple, she was unable to decide when the pile was high enough.

Maximum changes in resistance are missing in her action (see paragraph 2.2.1 "They Take off – But How"?).

For those who are perceptually disordered, the lack of tactile-kinesthetic information also explains their difficulty with toilet training. It can take a long time before perceptually disordered children and adults know when to empty their bladders. In order to empty the bladder adequately, it must be done before the pressure is maximal. Maximum pressure is only established when the bladder is filled. Those who are perceptually disordered need that kind of information telling them they now can empty their bladder, e.g. change the total resistance to no resistance. When this occurs, it has to happen in the shortest time possible in order to be perceived. This does not allow for a detour, such as, going to the bathroom.

To repeat: Children and adults who are perceptually disordered search for tactile-kinesthetic information by eliciting maximum changes in resistance. When there is no possibility to elicit such changes, the desired performance cannot occur. When maximum changes in resistance can be elicited, the performance can be successful. These two latter points are so essential to the learning of those who are perceptually disordered that it will be helpful to illustrate their meaning in two more examples.

Mrs. N. is brain damaged. She is a housewife. With the help of therapy, she should be able to again prepare simple meals for her family, e.g., breakfast or Birchermuesli (a Swiss recipe). For 6 weeks, the therapist has tried to teach her how to peel apples and how to spread butter on bread – all without success. The therapist tells me about Mrs. N.'s problems. I think about them. How long could we continue to peel an apple? The apple is round, and we could peel and peel until hardly anything is left in our hand. How long could we spread butter on bread? We could spread and spread for a long, long time. When peeling and spreading, we elicit only minimum changes in resistance. This provides enough information for a normal person to decide when the apple is peeled or the butter is spread. However, the changes in resistance are too weak and do not give those who are perceptually disordered the information they need to reach such a decision.

Perhaps this explains Mrs. N.'s difficulties and lack of learning. Could we change the situation? We try. Instead of peeling, we begin by cutting the apple. We divide the apple in half; then we cut each half in two to get fourths. We continue dividing until there are many little pieces of apple on the table. Only now do we begin to take the skin off with a chop of the knife – one piece of apple after the other. With the first piece, we cut once and the skin is off. With the next piece, we cut once and the skin is off. Mrs. N. is beaming and exclaims, "I can do it"! At the end there are many pieces of apple, all peeled, on the table. The preparation of Birchermuesli can be continued.

For spreading butter on bread, we put the slice of bread on the cutting board and cut it until there are many little pieces of bread on the board. A small piece of butter is also cut with the knife and pressed onto one of the pieces of bread, and it's finished. Then we cut another piece of butter, press it onto another piece of bread, and it's finished. We continue in this way until several little pieces of bread, each one with some butter on it, are on the plate. This will be breakfast for the family. Mrs. N. beams and again exclaims, "I can do it"!

Why is it that Mrs. N. can now perform? How has information changed when we have her go from peeling to cutting and from spreading to pressing? Each action of cutting and pressing is short in duration. Together with the change in resistance elicited, the information becomes decisive. Each cut and each pressure in this *short time span consists of a maximum change in resistance:* First, the knife moves through the air toward the apple; there is no limitation to the movement. Then the knife cuts through the apple until it is totally blocked by the cutting board. The same is true for the pressure of the butter on the pieces of bread.

The possibility of eliciting maximum changes in resistance in a short time span was missing for Mrs. N. in the previous peeling and spreading situations. Perhaps you would say, "But I can look at the apple and see if it is peeled? I can also look at the bread and see if the butter is spread on it". Of course, you and I as normal persons can refer to such visual information to decide if a movement has reached its goal or not. Mrs. N. can also look. She sees and can grasp the apple or bread. She obviously depends on the maximum changes in resistance to give her adequate information when she is performing a specific activity with them to get a desired effect. She needs tactile-kinesthetic information to know when the movement begins, how it should be continued, and when it is ended. The visual information of that situation did not allow for such decisions. We will return to this point in the next section.

The need for maximum changes in resistance in order to perform is also demonstrated in the next example of a child with a perceptual disturbance.

We have already described how K., 9 years, could not drink from a paper cup and was unable to eat a cookie (p. 122). In this example, she is still in the special home. Once a year the children get the opportunity to have some fun with pumpkins. They scoop out the pumpkin seeds, carve faces or other designs in the rind, put candles inside, and light them in the evening. Adult visitors help the children individually at the home. I am responsible for K. and it is obvious she cannot perform the task by herself.

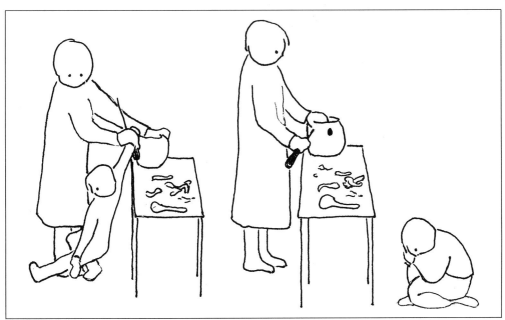

I begin to guide K.'s hands in grasping the pumpkin. As soon as we touch it, her whole body stiffens and jerks away. When I try to cut through the pumpkin, guiding her hands, I can no longer control her body. I let her go and finish that part of the task myself. I try to guide her again to remove the soft pulp and seeds, but she reacts with the same jerky, wild movements.

Again I continue without her and cut some star and moon shaped openings in the rind. Then I take her fingers and guide them to explore the openings.

When she first touches the pumpkin, she reacts with jerky movements. However, as soon as her finger enters the opening of the pumpkin, she stops those movements. Her face expresses a special tenseness which remains throughout the exploration. Movements become smoother and her tone, more adapted.

I can guide her easily now. We first explore one opening, and then a second, third, and a fourth. I can feel more and more how she begins to take over the movement. Finally, I can let go of her hands, and K.'s fingers move spontaneously towards the pumpkin to feel its openings.

What has happened?

Let us analyze the conditions of resistance which block K.'s movements. When her arms are guided through free space, no change in resistance is elicited between her arm and the surroundings. Then as she touches the cool pumpkin, she glides along its hard surface. The changes in resistance elicited by such movements do not vary. Then comes the hollowing out of the pumpkin, the penetrating into the soft pulp. The changes in resistance are small and do not vary much. But when we insert her finger into the opening, it is just big enough for one finger to squeeze in. The restricted space of the opening blocks the side movements of the finger with total resistance, and finger movement is therefore highly restricted. There are only two possibilities: Push the finger through the opening or pull it out. In other words, when her finger is inserted into the opening which offers total resistance all around, we elicit a maximum change in resistance. Thanks to this source of tactile-kinesthetic information, K. is able to take over the exploratory movements.

To Summarize:

Our observations indicate that children and adults who are perceptually disordered search for information. They are not different from normal persons in this regard.
However, children who are perceptually disordered attempt to elicit visual and auditory effects more frequently and at an older age than normally developing children.
In addition, it is startling to note that when perceptually disordered persons search for information, they elicit maximum changes in resistance within the shortest time span possible. It appears that those who have a severe perceptual disorder cannot use minimal changes in resistance.

3.2 When Information Is Deviant

In Part I of the book we described how normal children discover the world through touch. These attempts at touching elicit an intensive interaction between their bodies and their surroundings in the form of changes in resistance. *Such tactile-kinesthetically received experience permits children to become familiar with their surroundings. Visual and auditory information does not appear to be essential for this interaction.* On a higher level of integration, and only in coordination with tactile-kinesthetic information, does visual and auditory information add to the interaction experience (see Part I A, paragraph 1.3 "I Perceive the World Around Me: I Embrace It"). The lack of tactile-kinesthetic information affects the interaction between the perceptually disordered and their surroundings. Due to not acquiring sufficient tactile-kinesthetic interaction experiences, their surroundings remain unfamiliar. Inadequate tactile-kinesthetic information results in an over-emphasis on visual and auditory information. It remains fragmented due to its lack of integration with adequate tactile-kinesthetic information. As a result, the information which those who are perceptually disordered, both children and adults, pick up in a given situation differs from the information perceived by normal children and adults. This deviancy of information explains several of the described peculiarities.

3.2.1 ...Then Problems Are Recognized But Not Solved...

When the behavior of perceptually disordered persons is observed in daily living situations, we notice that they receive visual information. Frequently, *the visual information allows them to recognize problems, but it appears not to provide them the necessary information to solve the problems*. Often, children and adults with perceptual disturbances can describe the problem verbally. Recall the examples of paragraph 1.2 "They Talk Incessantly". B. noticed that one of the toys fell onto the floor. She said, "Oh, it fell down. Please, pick it up"! but does not move to get the toy herself. C. saw that his shoes were laced and asks someone to untie them. He made no effort to do it himself. Mr. F. and Mr. K. recognized they were missing some things. They asked others to get this or that for them. They do not appear to make any effort to get the things themselves. Is this laziness? Or is it perhaps that those who are perceptually disordered are able to see and can verbally name what they see, but that this is insufficient information for them to solve the problem? To answer this question, we observed how those with perceptual disturbance behave when they attempt to solve a problem in an actual situation. In paragraph 2.1.1 "They Withdraw from Touching, Become Tense and Look Away" we described the behavior of children with perceptual disturbance when they encounter obstacles while walking. D. is supposed to walk down a hill in order to reach some stairs. He begins to walk downward. What is startling is that he closes his eyes and thus excludes visual information; he appears to direct his full attention to feeling. We can infer that the tactile-kinesthetic information helps him to solve the problem, but then he has difficulty receiving enough tactile-kinesthetic information and consequently he closes his eyes. This observation suggests that in many situations perceptually disordered persons do not succeed in the search for sufficient tactile-kinesthetic information. This could be the case for B. who does not move to pick up her toy from the floor (see paragraph 1.2 "They Talk Incessantly").

Besides the difficulty of solving problems which are recognized, there is another difficulty.

3.2.2 ...and the Surroundings Are Still Unfamiliar

When we are in an unfamiliar situation, a few moments are needed to evaluate the situation. The amount of time needed obviously depends on how complex it is and demands the processing of visual and tactile-kinesthetic information. Stored interaction experiences help to order that information. For visually oriented persons who are perceptually disordered, such situations are difficult to evaluate. It becomes evident again *that visual information helps them to realize that the situation is a new one, but it does not provide them with the possibility of becoming familiar with it*. This requires tactile-kinesthetic information.

The search for tactile-kinesthetic information can often be clearly observed. At ages 10, K. moves along the walls of the new gymnasium. At ages 11, K. sits on the floor in the hallway of the new school building and no longer moves. Mrs. L. constructs a niche in one corner of her room using the walls, the bed, and a nearby table. These examples remind us that in addition to visual information those with perceptual disorders try to get as much tactile-kinesthetic information as possible (see paragraph 2.1.2 "They Know About the Rules of the Stable Support and the Side"). Similarly, we can interpret the observations we have described in relation to human contact. C., 12 years; K., 11 years; and E., 14 years, need to touch the persons with whom they are talking.

The examples suggest that persons with perceptual disorders process visual information, but it is not enough for them to be able to behave appropriately in different situations. The difficulty, or even impossibility, of picking up the tactile-kinesthetic information within situations keeps their surroundings more or less unfamiliar to them.

The use of visual information exclusive of the corresponding tactile-kinesthetic information about the event also explains other peculiarities.

When normally developing children see a peeled banana, they know that when touched, it will be moist, slippery, and sticky. Somehow they succeed in receiving the visual information to retrieve the corresponding visual and tactile-kinesthetic experience. But this is not so for those who are perceptually disordered. They also see the banana, but the visual information does not prepare them for what they will feel when touching the banana. In other words, they look at the banana and touch it, but then jerk away. The touching obviously consists of unfamiliar information. We have already mentioned this jerking away and getting tense when touching something (see paragraph 2.1.1 "They Withdraw from Touching, Become Tense and Look Away"). This is especially so when they touch something moist, slippery, or coarse. It can also be noticed when a support is unstable, as in the case of B., 5 years, whose therapist wanted him to sit on the big ball and he refused to do it (see paragraph 2.1.2 "They Know About the Rules of the Stable Support and the Side").

It is interesting to note that there are always situations in which no maximum changes in resistance are elicited. D. tried to make the tuna fall from the spoon into the tomato. He became very tense while doing this. Such tenseness is in direct contrast to the situation in which he was able to press the tuna into the tomato with force. Because he was pressing against a support, his tenseness decreased.

The deviancy of information received which results from the lack of tactile-kinesthetic interaction experiences, causes other difficulties for those who are perceptually disordered.

3.2.3 They Hardly Know What Is Happening Around Them

Several peculiarities were described in paragraph 1.4 "They Are Labeled Ill-Mannered". M. was considered impolite when he pushed through the group of customers to reach the cashier. Children at the school for those who are perceptually disordered held onto the railing when they went down the stairs and didn't notice the older people in the same situation who also needed to hold on to the railing. T. used his elbows to make a path through the group of tourists who blocked his way.

These kinds of situations resemble each other. In each one, the perceptually disordered children were busy solving a problem: M. figured out how much he had to pay and was ready to pay it; the children on the stairs had to concentrate on going down without falling; and T., on his way to shop, was concentrating on finding his way but came upon an unexpected crowd of people.

We have already mentioned that when solving problems those who are perceptually disordered extract mainly visual information from a situation, due to the difficulty of receiving enough tactile-kinesthetic information. Since they must concentrate fully on picking up tactile-kinesthetic information in order to solve a problem, it is difficult for them to collect enough information about what other people around them are doing during the same period of time.

But there is yet another even larger difficulty. *To be able to represent the problems of other people demands an extended amount of stored tactile-kinesthetic interaction experiences as well as the possibility to retrieve them.* Here one can observe the failure of those who are perceptually disordered. It is the lack of representation – not the lack of education – which causes those who are perceptually disordered to fail in social situations.

This also explains the difficulty of teaching M. that he cannot simply put any key he sees into his pocket. He knows he can open doors with a key, but he cannot represent that a specific key belongs to another person who may be elsewhere at the moment, and that this person uses the key to open her/his car or her/his house.

The lack of tactile-kinesthetic interaction experiences also explains the peculiarities which were described in paragraph 2.2.4 "When Only the Moment Exists".... Some perceptually disordered children fear the dark much more frequently than normal children. They also do not readily recognize danger. It is assumed that they cannot represent what is going on in the environment.

The fact that children who are perceptually disordered can hardly wait is related to the difficulty of representing events in the environment and is caused by the lack of tactile-kinesthetic interaction experiences. In the situation at the railroad station, B. knew the train would be leaving and could represent the action of it leaving, but he could not represent the activities of the station master that would occur up until that time. Similarly, he couldn't represent the activities of the conductor and the engine master. The verbal explanations by his father couldn't provide him with the needed representation, nor could they compensate for the lack of tactile-kinesthetic interaction experiences which are a prerequisite for such representation.

In paragraph 2.2.5 we described situations in which "causative actions do not correspond to the situations". In order to adjust causative actions to situations, we must be able to internally represent and anticipate the cause-effect event *before* the event happens. And this, as already noted, is very difficult, if not impossible, for those who are perceptually disordered.

There is a fourth peculiarity which has troublesome consequences for the development of children who are perceptually disordered.

3.2.4 The World Does Not Become a Surrounding World

Let us return our thoughts to children who are normal, as described in Part I A, paragraph 1.4.4 "The Multitudinous Ways of Touching and Releasing". It begins with the example of T., 7 months.

The situation: T. is lying on her stomach on a blanket on the floor and objects are scattered all around her. The objects do not touch each other, but they are all on the same support as she is.

The problem: T. discovers these objects. She tries to touch them – first this one and then that one. How does she get to these objects?

Solving the problem: She has to move the arms – including the hands and fingers –

and the trunk. Each movement depends on how far away on the support the object is from her body. The support is involved with each movement.

Let us observe the movements of the *hands*. We can see how T. changes hands, while at the same time, coordinating the activity of them. In the first picture, each hand holds an object. In the next, the left hand presses the sieve down on the support and the right hand lets go of the straw saucer. In the third picture, both hands are occupied with the same object. It is impressive to watch the difference in touching movements between the two hands and how this movement changes from one picture to the next. In the fifth picture, T. uses the right hand to reach out (in the first picture it was the left one) and to explore the straw saucer. Both hands are now on the sieve in the seventh picture. Let us compare it with the fourth one: the hand holding the object is first the left and then the right. What a wealth of tactile-kinesthetic information! Even from such a wealth of information, though, one can discuss only fragmented pieces of information.

Let us consider the *changes in resistance* as the most important information received through the tactile-kinesthetic system. In the first picture, T. reaches for the sieve with her left arm. This arm is on the support, and T. can feel its stability. Now the fingers touch the sieve and press it down on the support. All the fingers feel the change in resistance elicited by this pressure. In this way all fingers receive similar information. In the second picture we see how T. holds the sieve. It gives way, and she explores taking it off in the sense of something "moving along an immovable support". She appears to be so impressed that the second hand is now part of the exploration. T.s *trunk* moves very little – just enough so she can be sure the support is a stable one and can thus function as a reference surface for the various bodily movements.

In this way T. gains important experiences with changes in resistance. They first occur between body and support, and then they involve body, object, and support. They allow to establish multiple kinds of relationships. The touching and the moving of objects is constantly related to the stability of the support as a reference. The support functions as a reference surface also for experiencing the coordination of the hands, the movements and coordination of the fingers, and the neighboring relationships between objects.

At this point we will leave the development of normal children and return to those who are *perceptually disordered*. It appears that children who are perceptually disordered develop in a different manner. The following example illustrates this:

F., 22 months and perceptually disordered, lies on her back most of the time. She cannot crawl. She cannot hold onto the sides of things to pull herself up, even when a chair is placed next to her. If an object is presented to her, she grasps it, shakes it, and hits the floor with it. If she is asked to release the object, she can seldom do it.

For a few days now she has been able to walk if someone stands her up on her legs, but then she is not able to sit down without help. Also, she cannot pick up an object from the floor.

She possesses the correct motor patterns, but she can hardly apply them. She learns rapidly in situations in which she is provided with tactile-kinesthetic information, but as soon as the situation changes, she can no longer repeat the movements she has just performed.

How will further development be for F.? On the basis of our observations and experiences with children who are perceptually disordered, we can assume the following:
F. will utilize more and more of her time for walking. She will increasingly turn to objects from that upright position. Consequently she will touch the surroundings from free space. This means that her arms and hands will move through the air and not across a support as normal children do

when first exploring their surroundings. What does this mean for the development of touching movements, and later, for the development of holding something and taking it off? Children who are perceptually disordered, like children who are normal, depend on maximum changes in resistance when touching and when being touched, but they do it for a much longer period of time.

When perceptually disordered children are finally ready to go and grasp what they see, they have already learned to walk. Since they now touch the objects they see in free space, and not on a support, they cannot elicit changes in resistance involving their hands, a support, and objects. Consequently, their movements of touching objects are not directed to elicit these kind of changes in resistance. They *cannot use the support as a reference surface* like children who are normal (see the example of T., p. 148). How will they reach a harmony of finger movements? A unity in the use of both hands?

Children who are perceptually disordered use *another kind of reference point* for their movements of touching. They *use the fingers.* One finger touches the object from one side and the other finger touches it from the other side. Now both fingers press firmly against one another, with the object in between, until resistance is total. In a short time span, this action results in a maximum change in resistance created by two fingers and the object.

The difference in the reference system leads to peculiarities which particularly involve the development of touching movements. This is shown by the ways they take something off a support and how they handle neighboring relationships. In paragraph 2.1.4 "...and Don't Succeed in Embracing Things," we described how children who are perceptually disordered use *peculiar finger movements* to grasp things. Instead of using several fingers, they touch and hold an object with just two fingers like one would hold with a pair of tweezers. When they touch a surface as R. did while cleaning the table, their fingers spread out. Recall also the strange finger movements of L. when she took seeds out of a grapefruit. In paragraph 2.1.3 "They Have Two Hands but Often Use Only One...," we described the peculiarity of perceptually disordered children who skillfully use only *one hand*, even at 10 to 12 years of age or older. The peculiar movements of the fingers and the use of only one hand are interpreted as being the result of an absent reference system involving a support.

The next peculiarity concerns taking off and, closely related to it, releasing (see paragraph 2.2.1 "They Take off – But How"?).

We have already described how children who are perceptually disordered grasp an object by approaching it from free space. When they have taken the object off, we can observe the next difficulty – releasing it. It is not like normal children have learned. For example, Ch., 4 years, 6 months, threw objects he had taken off through the air instead of releasing them on the support. This peculiarity of throwing can be explained by recognizing it as a lack of experience with using the support as a reference system when exploring the surroundings.

A third group of peculiarities is related to neighboring relationships. In paragraph 2.2.2 "And Where Is the Neighborhood"? we raised the question: Where are the objects in the surroundings which I can see in relation to my body and in relation to each other?

Visual information alone is not sufficient to judge distances between our own body and objects. Therefore, children who are perceptually disordered can hardly judge distances, if at all. R. became panicky while hanging on rings and his feet lost contact with the floor. Consequently, he had the impression of being close to the ceiling. His difficulty of judging how near his feet were to the floor was obvious. D. crawled on the ground underneath a rope even though there was enough space between his body and the rope. J. stopped and prepared to step over the door sill each time he saw it, even though it was still far away. S. tried over and over again to reach for a basket of nuts. He went around the table several times and still was unable to judge that the basket was too far away. The example of T. with the dog (see paragraph 2.2.4 "When Only the Moment Exists"...) further underscores the difficulty that those who are perceptually disordered have with judging distances. T. was still quite some distance from the house with the dog he feared. He saw the house but could not realize the difference between being close and being far away. He feared the dog would come right out and bite him. In paragraph 2.1.2 "They Know About the Rules of the Stable Support and the Sides," we referred to K. who could not walk over the rain gutter.

However, the fact that visual information alone does not provide enough information to judge distances is not the only problem; visual information can even lead to *faulty judgments*. The next example illustrates this aspect.

C., 11 years and perceptually disordered, is making fruit juice. To do this he uses blackberries that are partially frozen. When he presses on them, his fingers become too cold, so he takes a clear glass bowl and presses it on the berries. The outside of the bowl becomes coated with the red juice. He lifts it to his mouth to taste the juice he sees "inside" the bowl, but nothing is there. C. is surprised. He examines the bowl thoroughly. Finally he turns it over and realizes that he can lick the juice from the bottom.

Children and adults who are perceptually disordered have difficulty perceiving the three dimensions of the world around them. This leads to the difficulty in differentiating between being in *front* of something from being in *back* of something. C. assumed that the juice he saw in the bottom of the bowl was in front of it, but actually it was in back of it. This difficulty prevents the world from becoming a surrounding world for those with perceptual disorders.

Perceptually disordered children appear to be able to judge the neighboring relationships involving body and support, or involving two objects, but only when they touch each other.

In paragraph 2.1.2 "They Know About the Rules of the Stable Support and the Sides," we mentioned R. who kept his feet in continuous contact with the floor when dancing. This is the same child who panicked while hanging on the rings in gym class and was not in contact with the ground. This means also, that they can judge that two objects are in a neighboring relationship when the evaluation of the relationship does not require a reference to the support. In paragraph 2.2.2 "And Where Is the Neighborhood"? we described how children who are perceptually disordered can construct neighboring relationships between two objects only when they touch each other. For example, C. inserted objects into gaps; K. connected water pipes; D. prepared tea by hanging the tea-ball on the pot; T. put the sausages on the dough; and D. hung the clothes on the clothesline.

It is in this way that children who are perceptually disordered construct a world which is different from the world of children who are normal, and which doesn't become a surrounding world. They acquire the knowledge that one may grasp, through free space, objects they see, but they do not know how far away they are. In their world the reference of objects they see to a support which connects them with their own bodies is missing.

Children who are perceptually disordered can become skilled in walking. The more skilled they become, the faster they run to grasp objects they see, approaching them from free space and releasing them again into free space. As a result, their behavior becomes more and more hectic, and finally they are labeled as being hyperactive. We described such behavior in paragraph 1.1 "They Are Either Too Hectic or Too Quiet".

Explanations of the peculiarities which can

be observed among those who are perceptually disordered would be incomplete without mentioning the notion of *capacity*. Capacity is strongly related to the search for information.

3.3 The Limitation of Capacity

The search for information requires capacity. Here capacity refers to the quantity of information which one can process within a given time span.

The capacity to perceive is limited. When we are intensely searching for information, we become fatigued. Researchers in the field of *information theory* have studied this problem. Broadbent (1958, 1971) and Miller (1967) described research findings which led them to assume that the capacity of the human nervous system to process input is limited.

3.3.1 What Are the Consequences of a Limited Capacity?

In ordinary daily living situations we hardly recognize the limitation of our capacity. As people who have developed normally, we organize our activity so that our capacity is sufficient for it. However, sometimes we are very tired, e.g., we wake up with a headache, an important co-worker is missing, or someone in the family suddenly becomes seriously ill. In such situations, we can experience the limitation of our capacity to process information. Perhaps we cry out, "That's enough! Please be quiet! Let me go! I can hardly think! I can barely breathe!

Performances deteriorate under stress. The deterioration is systematic and its effects are predictable (Cherry, 1957; Affolter, 1970; Affolter & Stricker, 1980).

Many peculiarities of children and adults who are perceptually disordered can be explained by such a deterioration of performance due to capacity limitations. In paragraph 2.2.2 "And Where Is the Neighborhood"? we stated that the construction of neighboring relationships can be adequate when it includes a direct touching of objects and when it is simple. Such situations included the following: inserting something into a gap, connecting one pipe to another or adding one more pipe, hanging a tea-ball on the pot, and putting a sausage on the dough. Perceptually disordered children, however, fail when they must consider *two relationships at the same time*. D. could not hang the tea-ball on and in the pot at the same time; T. could not put the sausage on and in the dough at the same time.

They also fail when they have to establish relationships involving more than *two objects at the same time*. D. could relate the clothespin to the piece of wash but could not figure out how to put it on the clothesline at the same time. A. could relate the cord of the vacuum cleaner to an opening in the wall but could not conceive that, at the same time, an electric source is related to the cord and an outlet in the wall. Neither could she consider a relationship involving the two packages and the shopping bag nor the relationship among all of these objects and her arm.

In paragraph 2.2.3 "The Sequence – When Something Is Missing or When You Cannot Go Back" we described difficulties which can also be explained by capacity limitation.

When children and adults with perceptual disorder make errors, *they often cannot reverse the error; they must start all over again.* When K. put the different sizes of disks in decreasing order on a stick, he made an error. He noticed the error only after he had finished inserting the next disk; he started all over again.

To correct an error within a sequence of actions is more difficult than to start all over again. To reverse the error, we must determine its location in the sequence. We must differentiate between the correct and incorrect steps. Where are they? What comes first? What comes afterward? Most of the time this overloads the capacity of those who are perceptually disordered, and they begin again.

In the same paragraph 2.2.3, we described how children and adults who are perceptually disordered fail when they are required to order several dependent activities. That is, the activity must be done in a certain series of steps. This is the problem of *serial ordering*, which has been discussed in previous publications (Affolter & Stricker, 1980). The following examples illustrate the difficulty which those who are perceptually disordered have with serial ordering.

B. has everything ready for pasting: paper, paste, and pictures. She even has the paste open. She knows she has to put the paste on the back of the picture at some time and put the picture on the paper at another time, but she cannot determine the exact sequence of doing all of this. What comes first? What next? What last? Likewise, Mr. R. has a similar difficulty with the mixer, the cord, and the plug. With one hand, he holds the mixer; with the other one, the cord. The extension cord is on the table. He knows he has to connect the cord to an extension cord and that he needs two hands. But how is he to get one hand free? Which action comes first? Which one follows? Which one is last? Mr. R. cannot perform such serial ordering. Another similar example is the description of R. with the lemon juicer. He does not succeed in ordering the three items – the upper and lower parts of the juicer and the location of the lemon. In other words, he cannot decide which part to take first, which one follows, and which one is last. He recognizes his difficulty and becomes panicky. He throws everything off the table. The second time, he takes the lower part and puts the lemon into it. For his last attempt, he puts the upper part of the juicer on top (see paragraph 1.3 "They Are Called Aggressive").

This difficulty in ordering activities into sequences also explains the difficulty that brain damaged patients have in calling the nurse to come and help them go to the bathroom during the night.

Such *complex situations are often simplified* by leaving out some steps. The first examples in paragraph 2.2.3 illustrated such a behavior. D. was cooking peas. She got the pan, put the peas into the pan, turned on the hot plate, but forgot to pour water into the pan. While drying his hair, T. sat at the right place under the dryer, correctly turned on the dryer, but forgot to take off his swim cap.

Another set of behaviors peculiar to those who are perceptually disordered can also be explained by a capacity limitation. When *touching something, they often turn their eyes away* (see paragraph 2.1.1 "They Withdraw from Touching, Become Tense and Look Away"). D. looked away when he touched the tuna or when he went down a hill. The more slippery the ground, the less C. looked at where he was walking. In each of these examples, children who are perceptually disordered concentrated on tactile-kinesthetic input. The looking away can be interpreted as a sign of limited capacity. They could not feel and see at the same time because their capacity was focused on perceiving the tactile-kinesthetic information needed to perform the task.

3.3.2 When I Can Order Information

Knowledge about capacity limitation has raised several questions. Among them is the question of whether capacity can develop. Observations have shown that as they grow older, children can store more information. Could we, therefore, explain such findings by assuming that capacity increases with age?

Miller and others in information theory offer another explanation. Since this explanation is important for understanding the nature of the difficulties of those who are perceptually disordered, we will briefly summarize their research findings.

Miller (1956) investigated how much information one can perceive. He observed that the amount of information is the same whether one perceives it through the visual, auditory, or tactile-kinesthetic sensory system. The amount is also the same from one adult to another. Miller called it the *magic number seven plus or minus two*. The plus-or-minus reference emphasizes the fact that there are situations when we perceive less information, e. g., when tired; in other situations, we perceive more information, e. g., when rested.

But what does the number seven refer to? Miller used the notion of *bundles*. We can put several different objects into a bundle. By making bundles we can carry more objects than when only single objects are carried in the hands. The better the objects are bundled, the more objects we can carry.

It is likewise with information. We can bundle it. A person who is skilled in bundling information can carry more information in the bundles than one who cannot bundle very well.

As children get older, the amount of information they can handle at one time can be explained in terms of more adequate bundling skills.

The ability to make bundles is a requirement for perceiving, and consequently, for storing a certain amount of information (Norman, 1982).

3.3.3 What Happens When I Am in Search of Tactile-Kinesthetic Information...

We described how children and adults who are perceptually disordered try to elicit maximum changes in resistance. We inferred that they are searching for tactile-kinesthetic information. We also described how such information helps them begin to solve problems and to evaluate the effects of their actions.

We further described how a lack of tactile-kinesthetic information causes a lack of interaction experience, and as a result, the world does not become a surrounding world. The lack of experience and the difficulty of perceiving enough information in an actual situation reinforce each other in their effects and make the bundling process difficult.

Usually, the capacity of those who are perceptually disordered is completely used in the search for information and in the bundling of it. As a result, capacity becomes overloaded, and since the bundling is poor, the storage of received information is affected. This is often incorrectly interpreted as a disorder of memory.

However, some children and adults who are perceptually disordered reach a certain level of *compensation*. They may compensate by taking more than the usual amount of *time* to perform. They perform very slowly. Parents and teachers of these children report that they need much more time for their homework than other children, or that they cannot follow when they have to write from dictation.

Other children and adults who are perceptually disordered try to *limit the amount of information* inherent in an actual situation so that their capacity does not become overloaded. They retire to a calm location where there are hardly any changes; they behave like outsiders; or they stay at one place and do not move (see paragraph 1.1 "They Are Either Too Hectic or Too Quiet").

Those who are perceptually disordered experience an overload of their capacity earlier and more frequently than normal people. When this occurs, they are conscious of their failure, and unlike those who are mentally retarded, they become *frustrated.*

F. was going to make a car. Instead, he only succeeds in putting one block on top of another and in wedging a button into the hole of the block. We can assume that he became conscious of his failure because his action became hectic. At the same time, he began to turn the button over and over again and talked incessantly. This behavior is an expression of high tension – tension which can be interpreted as a sign of frustration (see paragraph 1.2 "They Talk Incessantly").

This kind of tension can increase until a panic state is reached. R. couldn't squeeze the lemon, so he threw everything on the floor. Ch. couldn't swim in the water, so he began to shout (see paragraph 1.3 "They Are Called Aggressive").

3.3.4 ...and the Competence Does Not Become Performance?

We repeat, the capacity of children and adults who are perceptually disordered is used up faster and overloaded sooner than it is in normal people. Those who are perceptually disordered must not only search for information more intensively than those who are not, but the information picked up is more often unfamiliar and incomplete.

Therefore, we must point out again that, *whenever there is a lack of information and the capacity is overloaded, performances deteriorate.* This is the case for all people, but it is more frequent and happens more quickly for those who are perceptually disordered.

In this regard, it becomes important to differentiate between a deterioration of performance due to the situation and one due to a general inability to perform. In other words we must differentiate between *competence* and *performance*.

I have certain rules available – rules of touching and rules of acting upon. This means that the rules are part of my *competence*. Children and adults who are perceptually disordered are also competent in such rules (see Chapter 2, "We Observe: They Have It and Yet They Don't Have It").

A prerequisite for a specific performance is always the competence for that performance. To be competent, however, is not the only prerequisite. Performances always occur in an actual situation. The complexity of the situation is another prerequisite for performance. The following example illustrates the interrelationship *of competence, performance, and situation*:

If I am able to play the flute, I am competent in playing it. This does not mean that I can play the flute in any situation. At home in familiar surroundings, I might not have any difficulty playing a specific piece of music, but just as soon as I go to my music lesson, my performance might be poorer. It will probably be worse when I give a recital in a concert hall where I face a large audience. In that circumstance, my performance could deteriorate.

Thus, the complexity of a situation influences the performance. Depending on the situation, the production of a performance is more or less difficult and might even break down. This is often the case for those who are perceptually disordered.

Considering the dependency of the situation and the description of the lack of tactile-kinesthetic information in perceptually disordered children and adults, it should be evident that they reach a level of performance less often than we do. Many more situations are unfamiliar to them than for those of us who are normal. Therefore, their competence leads less often to a performance than it does for us.

As a result, when those who are perceptually disordered cannot perform in an actual situation, we must be careful not to interpret it as a lack of competence. It is much more probable that competence is present, but *performance has deteriorated* in the particular situation. To investigate the situation and its familiarity for the one who is perceptually disordered, we must consider the amount of information inherent in that situation. Perhaps the situation offers little tactile-kinesthetic information which is of the utmost importance for solving problems. If it is possible, and it usually is, we must try to provide tactile-kinesthetic information when it is established that it is lacking. How can we provide this kind of information? These problems will be discussed in Part III, "Learning in a Wirklichkeit.

To Summarize:

Children and adults who are perceptually disordered fail in daily life. We described their difficulties and the way people in the environment often misjudge them. The people around them often become tense and actually cause an increase in their failures. Unfortunately, it is impossible in the context of this book to more thoroughly describe the fears, worries, and questions that families have if failure continues for their child or adult member who is perceptually disordered.

We observed how perceptually disordered children and adults try to apply the rules of touching and acting upon in their search for information about changes in resistance and how *a performance can be produced in a specific situation but can rapidly deteriorate when that situation changes.*

We observed the specific difficulties they have in establishing relationships, in ordering sequences of actions and events in time, and finally, how all of these problems cause a failure of living in a Wirklichkeit.

Step by step, we considered what it means to search for information, why it is *necessary to have tactile-kinesthetic information,* and what happens when those who are perceptually disordered do not receive the necessary tactile-kinesthetic information. We also considered how the world does not become a surrounding world for them, i.e., does not become a Wirklichkeit, and how, as a further consequence, their capacity to deal with information input overloads rapidly.

What can we do? We will reflect on this problem in the next part of the book.

Part III
Learning in a Wirklichkeit

Daily living happens in a Wirklichkeit
full of problems.
As a child, I try to discover
and to solve such problems.

Solving problems of daily living
creates difficulties
which I try to overcome by using my touch
to explore whatever is around me.

I gain knowledge
about the properties of things and
about neighboring relationships.
In doing this I discover myself.

I learn to act upon my surroundings;
I involve my own body
to elicit changes
and use my touch to explore them.

Much thinking has been done about the first steps of development of normal children. We pointed out the importance of tactile-kinesthetic information when interacting for normal development.

Using many examples of daily living situations, we also described the failures of children and adults who are perceptually disordered. We noted that their difficulties in receiving adequate tactile-kinesthetic information is the basis for the their conspicuous behavioral peculiarities.

We will now reflect on how we can help those people who are perceptually disordered.

- *In Section A* we will briefly discuss what is viewed as the *root or origin* of development and its importance for work with perceptually disordered persons. Although a number of questions are related to this issue, this discussion will focus on just a few aspects. For more details, the reader will be directed to the corresponding literature.
- *In Section B* we will discuss *learning through tactile- kinesthetic experience* by presenting practical examples. We will reflect upon the importance of tactile-kinesthetic input as it occurs during the solving of daily problems. Is it possible to *change behavior* because of *exposure to tactile-kinesthetic information*? That is to say, can one learn this way? How can I help perceptually disordered children and adults receive better tactile-kinesthetic information?
- *In Section C* we will describe the *beginning of production and representation.* These descriptions will complete the circle of problems – those of perception, daily life interaction, and language.

The presentations correspond to our knowledge at the *present time*. There is yet much to be learned, as we are still *progressing* up a steep mountain.

The examples will be mixed. Sometimes we will refer to normal children or adults and at other times to those who are perceptually disordered.

We hope you accompany us with understanding when reading through this part of the book!

A. Problem Solving Events Are the Root of Development

1 Development Occurs with a Surprising Regularity

Developmental progression appears in children with astonishing regularity. There is a wealth of literature available to support this observation. The results are universal. The following descriptions apply to all children everywhere in the world:
Babies perceive. During development their perception is extended (Salapatek & Cohen, 1986), and perceptual activity becomes more organized (Piaget, 1961). During the first year of life, children begin by imitating hand and body movements and then they imitate sounds (Piaget, 1962). At about 12 months we can observe behavioral characteristics which exhibit the child's knowledge of an object as being permanent – object permanency (Piaget, 1950). At the beginning of the second year, they show detour and means-end behavior (Piaget, 1952). The middle of the second year is characterized by the emergence of semiotic behavior which includes deferred imitation and the discovery of language (Piaget, 1962). The development of speech sounds includes a stage of babbling and a stage of reduced babbling. This latter stage occurs just before the beginning of phonological development (Jakobson, 1969).
The development of language includes one-word sentences followed by two-word sentences. This is followed by three-word sentences and finally by more complex sentences. Sentences with passive structure appear last and only at about 14 years of age (Menyuk, 1971). Consequently, the acquisition of language takes about 12 years, 6 months, beginning at 18 months and lasting until 14 years (Palermo & Molfese, 1972).
Children who are profoundly deaf do not differ from those children who hear in regard to the sequential appearances of different developmental performances.
Sensorimotor performances appear in the same sequence as long as they do not require auditory processing. For instance, deaf children imitate body movements at the same time as hearing children. Depending on the amount of hearing loss, the imitation of certain sounds will be missing. Semiotic performances, e.g., deferred imitation and symbolic play, begin at the same time for deaf children as for hearing children. They interiorize the cognitive and emotional content of their experiences. They reach the stage of reversibility and acquire performances of concrete intelligence in the same sequence as do hearing children (Affolter, 1954, 1968, 1985; Affolter & Bischofberger, 1982). Furth (1966) compared formal-logical performances of deaf children and adults with those of hearing ones and found no significant differences. The differences which he observed corresponded to the differences found when comparing children from rural regions with those from cities.
Bischofberger (1989) summarized research findings on the development of blind children and complemented them with his own studies. He reached the conclusion that *the development of blind children is basically the same as that of seeing children.*
When these kinds of research findings are considered, one can conclude that *neither vision, hearing, nor language are necessary conditions for normal development.*

2 But What Happens When Children Fail in Perception?

For 15 years we observed the development of *children with different kinds of perceptual disorders*. (For the description of different disorders of perception, see Affolter, 1974a, 1976, 1977; Affolter, Brubaker & Bischofberger, 1974.) These observations were compared with the development of children who were normal and with those who were hearing impaired. For a time span of 10 years, this comparative research was sponsored by the Swiss National Science Foundation (see list in introduction). Several publications present results of that research (see references Affolter and Affolter et. al., Annex).

The research findings emphasize deviations in development for children who are perceptually disordered. Among them are those who fail in tactile-kinesthetic input, or in the coordination of sensory modalities, or in the integration of temporal-successive stimuli.

Among other observations, the deviations are observable in the development of perceptual performances, in the sequence of developmental performances, and in problem solving activities.

2.1 The Development of Perceptual Performances Is Deviant

Normal children were tested on perceptual performances. As they increased in age, we observed an improvement in the recognition of successive patterns and forms under different modality conditions (Affolter & Stricker, 1980, pp.140-157).

Hearing impaired children did not differ from those who could hear in visual and tactile performances (Affolter & Stricker, 1980, pp. 23-29; pp. 39-66).

Blind children needed more time for auditory and tactile performances to improve (Affolter & Stricker, 1980, pp. 23-29; pp. 39-66), that is, they reached the same level of performance as seeing children but with a time delay (Bischofberger, 1989).

Children with *perceptual disorders* in tactile processing, or intermodal connections, or temporal-successive integration (we referred to them as perceptually disordered in this book) differed from normal, hearing impaired, and blind children. They did not perform like younger, normal children, their performances were *deviant*. They failed in the recognition of successive patterns in different sensory modalities. The more complex the patterns, the more pronounced were their difficulties. They differentiated forms presented in the visual modality, differing only slightly from those of normal children. However, when the forms were presented in the tactile modality, the differences were pronounced.

2.2 Developmental Performances Appear in a Different Sequence

Several findings point out that developmental performances of perceptually disordered children do not appear in the same sequence as in normal children.

Piaget described the development of normal children in his numerous books. For an exact description of performances mentioned here, refer to his books (Piaget, 1962, 1952).

Normal children begin with *direct imitation* a few months before they discover language. This is not the case for many children who are perceptually disordered. Some begin to comprehend language before they imitate (Affolter & Stricker 1980).

When normal children develop *language*, they begin with one-word sentences, seldom with longer ones. Afterward, two-word sentences are learned, and later, more complex ones. Among those who are perceptually disordered, some correctly produce longer sentences right away. After some time, though, the performance breaks down and two-word sentences begin.

Normal children *babble* before they can produce speech sounds. Perceptually disordered children usually do not go through a babbling stage. Nevertheless, most of them learn to speak.

At first, normal children *scribble*; then they begin to draw *human figures* by producing global shapes, such as a round head and "stick" arms and legs projecting directly from the head. Later on, they draw them in more detail. Drawings with constructions in *perspective* are only observable at later ages (Piaget & Inhelder, 1956). Among the perceptually disordered children, though, some begin with drawings in perspective, and only later on do the drawings of human figures appear.

These, and many other observations too numerous to mention here, support *different developmental sequence of performances* among perceptually disordered children, (i.e., between subgroups) and between the group of perceptually disordered children and the group of normal children. These findings allow us to assume that developmental performances *do not show a direct relationship* (Affolter, 1985).

2.3 Problem Solving Activities Are Deviant

Let us observe some activities shown by children who are solving a problem. A ball has rolled under a dresser. A boy tries to get it. How does he proceed? What kind of movements does he perform? What subgoals determine his various actions?

The boy has to explore how far away the ball is. Can he reach it by extending his arm? He makes a *hypothesis*: Yes, the ball is near enough. He tries to grasp the ball with his hand, an activity which also serves for *picking up some information*. His hand may grasp into empty space. He feels that; he receives *feedback*. The child *evaluates* this information and *concludes*: "The ball is too far away. I can't grasp it with my hands. What now"? He has to make another hypothesis: "Perhaps my mother can get the ball; or perhaps I can get it with a broom; or".... He *decides* to call his mother.

We can interpret the different actions observable during the solving of a problem as the expression of different processes: formulating hypotheses, picking up information, evaluating feedback, drawing conclusions, and making decisions.

We examined such problem solving processes in *normal* children at different age levels (Affolter, 1985). Older children scored significantly higher than younger ones.

Results of perceptually disordered children were significantly different from normal children of their own age and younger. It appeared that these children did not merely present a delayed development but also exhibited a deviant one. The most noticeable difference was found in the processes which were used to pick up information. Differences were smaller in processes related to formulating hypotheses, evaluating feedback, and drawing conclusions (Affolter, 1985).

3 How Can We Represent Development?

3.1 Interpreting the Results

The fact that development occurs so regularly in normal children and with corresponding regularity in hearing impaired and blind children forces us to search for an *origin* of development that does not depend exclusively on seeing, hearing, or language, i.e., an origin that would be the *same for all of these children*.

Perceptually *disordered children* differ from normal, hearing impaired, or blind children; they show a different sequence of developmental performances. It is assumed that *perceptual disorders affect the common origin of development* – the source.

In Part I, while searching to identify this origin of development, we described the behavior of babies and infants. They are in continual motion. We differentiated several patterns of action: touching, embracing, taking off, and acting upon. As children progress in development, these action patterns include interrelationships of increasing complexity between the child and the surroundings, until finally, they become embedded in sequences of goal-oriented activities which are directed at solving the problems in *daily living situations.*

Hearing impaired children are similar to those without a hearing impairment in understanding, and later on in producing problem solving events in daily life. The same is true of blind children, though they seem to need more time before they acquire a critical amount of tactile-kinesthetic experience in problem solving events in daily living (Fraiberg, 1977; Bischofberger, 1989). In different degrees, perceptually disordered children fail to solve problems in daily living situations (see previous paragraph, 2.3 "Problem Solving Activities Are Deviant"). Observations suggest that they fail to solve problems because the necessary amount of tactile-kinesthetic information in the actual situation is not perceived.

In short, these findings lead to the conclusion that *the origin of development can be described as a tactile- kinesthetic interaction which occurs between children and their surroundings in the form of problem solving events.* Furthermore, the deviations in the development of perceptually disordered children lead to the conclusion that they are disordered at the level of the developmental root.

These conclusions support a model of development which differs from the usual ones. What follows will be a brief description of this model.

3.2 The Model of Development

At the center of the model are the problem solving events in daily living. Tactile-kinesthetic experience as a result of the interaction between the child and the environment in the form of such events constitutes the *root* and stem of development. Similar to the growth of a plant, the root and stem of development expand with the extension of the tactile-kinesthetic experience in interacting with the environment.

In the course of development, children's earliest tactile-kinesthetic experiences become organized by the rules of touching, i.e., the rules of the support and the rules of the sides. (These experiences were described in Part I of the book.) Touching leads to embracing, taking off, bringing back, and releasing, always in reference to the stable support. Interaction experience extends and children begin to become familiar with the qualities of their surroundings. The tactile-kinesthetic experiences increasingly include more cause and effect events and become organized by the rules of acting upon, i.e., the rules of taking off and the rules of neighboring relationships. The children begin to know about the Wirklichkeit.

Toward the end of the sensorimotor stage, comprehension of the changes in the surroundings is so extensive that children begin to apply the rules of touching and acting upon, not only to explore what the Wirklichkeit is, but also to solve problems which arise in daily living. By doing this, they themselves start to produce changes in their Wirklichkeit.

Out of the stem branches begin to grow. Tactile-kinesthetic information related to problem solving in daily living is increasingly co-ordinated with visual and auditory information about the same events. Such tactile-kinesthetic information and its coordination with information from other sensory modalities are ordered in temporal sequences within the problem solving events – serial organization (Lashley, 1951). In this way, *perceptual performances* develop in direct dependency on the tactile-kinesthetic experience of interaction in the form of the problem solving events of daily living.

Sensorimotor performances, e.g., direct imitation, signal behavior, and permanency of objects (Piaget, 1962, 1952) develop due to the expanding experience with problem solving events. However, in order to develop, they require a greater amount of interaction experience than do perceptual performances.

As tactile-kinesthetic experience with problem solving events increases, it begins to be interiorized. The increase and the interiorization allow for the discovery of the *semiotic function* around the middle of the second year (Piaget, 1962). With further interaction experiences, semiotic performances continue to develop. According to Piaget (1962), language is one of several expressions of the semiotic function.

The stronger the root and the stem, the more perceptual performances improve and the more complex sensorimotor and semiotic performances become. For normal children around 6 years of age, the amount of tactile-kinesthetic experience with problem solving events is so rich and expanded that it allows for the appearance of performances at another level – the level of *concrete intelligence* (Piaget, 1950). Performances of concrete intelligence, therefore, demand a higher amount of tactile-kinesthetic interaction experience with problem solving events in daily life (including their interiorization) than performances of sensorimotor intelligence.

An even higher level of interiorization of tactile-kinesthetic interaction in problem solving events is necessary to reach the level of *formal-logical* intelligence (Piaget, 1950), which appears at about 12 years.

It is important to emphasize that performances at different levels of development do not depend on each other. All the developmental levels depend *directly on the root* in the same way that the branches of a tree do not depend on each other, but rather on the stem and the root. Thus, imitation as a characteristic of the sensorimotor level is not a prerequisite for the discovery of language (Affolter & Stricker, 1980), but rather, both imitation and language depend in a direct way on the stem and the root, i.e., they depend on tactile-kinesthetic experiences with problem solving events (see Part III C "Tactile-Kinesthetic Experiences with Solving Problems of Daily Living Are Interiorized").

The observation that hearing impaired and blind children, like normal children, are involved in tactile-kinesthetic interaction in the form of daily problem solving events, explains why there is no difference in their sequences of development.

Therefore, it is assumed that the appearance of performances at new developmental stages depends directly on the stem and the root. In this way, the model contrasts with the usual models of development. The usual models assume there is a direct relationship between the more elementary and more complex performances and the successive stages of development.

4 We Cannot Simply Wait

We stated that the sequences of developmental performances in perceptually disordered children are *different* from those of normal children. Therefore, we cannot simply wait, hoping that with time perceptually disordered children will gain enough experience with problem solving events in daily living to enable them to progress in development as normal children do. We must *intervene*. But how?

The most frequent answer given is: by *practicing skills*. This kind of intervention directs attention to some of the performances most often failed by perceptually disordered children and adults. For instance, they must be taught how to get dressed, how to button their coats, or how to hold a spoon when eating. They are taught how to draw a human figure, how to write a letter, etc. Disordered perceptual performances are practiced in a similar way by requiring the discrimination of forms through touching, seeing, or hearing. Eye contact and imitative behavior are rewarded when they occur, and the children are required to sit still and look at the teacher when they are being shown what to do.

4.1 We Should Not Practice Skills...

Certainly, we can gain skill by practicing. One can practice and become adept in knitting, cooking, sawing wood, typing, or using chopsticks to eat in a Chinese restaurant. *One can also teach skills to those who are perceptually disordered.* This may not even be very difficult to do. Some perceptually disordered children become very skillful all by themselves. Among the skills are juggling where round objects are thrown into the air and caught with impressive ease, and swimming where one can jump into a pool and move around without ever having swimming lessons.

However, the *practicing of skills* - even in great number - *cannot make perceptually disordered people progress in development.* This fact creates frustration for us when we consider how much effort is devoted to such tasks by those in the environment as well as by perceptually disordered persons themselves. There is so little success. As soon as one who is perceptually disordered is in an unfamiliar situation, the skills that have been learned break down. This indicates that the basic difficulties continue to persist in spite of all the practicing.

4.2 ...but Should Begin with "Problem Solving Events" and Mediate the Corresponding Tactile-Kinesthetic Information

In the previous sections, we described a developmental model that depicted tactile-kinesthetic experience with solving problems in daily living situations as the common root of different developmental stages. When a plant has *sick roots*, it will not become healthy when we treat the leaves or the branches. Similarly, it does not help when we practice skills with perceptually disordered children and adults. We must work at the root. We must treat the cause and not the symptoms.

The root of development is based on the problem solving events of daily living. These events must be the thrust of our treatment orientation. While working with the perceptually disordered, therefore, it is essential to include their *daily lives* and consider what *problems* could arise out of their individual life situations. We need to help them find ways to *solve* the problems particular to them. During these problem solving events, we must make sure they get enough *tactile-kinesthetic information* for interacting with the environment.

This means that one cannot turn to games or educational material to help fulfill this task. Games and educational material are not part of the Wirklichkeit. They demand pseudo-interactions and can be fun for people who want to relax and get away from their innumerable daily activities, but they are neither for children who hunger for experiences in tactile-kinesthetic interaction, nor for perceptually disordered adults who would like to progress.

Furthermore, in order to be of real help, we need to be aware of *what makes the root sick.* In Part II of this book, we emphasized that perceptual disorders cause a lag, and even a *deviancy*, in *information input*. This problem must be attacked when working with people who are perceptually disordered. The clinical work has to be integrated into the events of solving problems which arise in their daily lives. It is our duty to assure, that in performing such tasks, *more adequate tactile-kinesthetic information* is received. The success of our whole work depends on the success of this particular aspect. How this can be achieved will be the subject of the following sections, B and C.

B. Problem Solving Events Can Be Felt

1 I Feel and Can Change My Behavior

Unfortunately, we have hardly any knowledge about the *physiological* functioning of the tactile-kinesthetic sensory system (Schiff & Foulke, 1982). Similarly, there is little knowledge about the *development* of tactile-kinesthetic performances. Literature in the field of perceptual development is restricted to the discussions of visual and auditory perception. If tactile-kinesthetic development is discussed at all, it is only a short paragraph but it is usually not even mentioned. Tactile-kinesthetic input and its relationship to development is hardly ever written about (Neisser, 1976; Pick, 1980).
For us, however, the question of eminent importance is: Can we learn when we receive tactile-kinesthetic information?

1.1 I Learn from Tactile-Kinesthetic Experience

When children are in their natural surroundings, they frequently encounter problems they cannot solve by themselves. When parents notice this situation, they often help their children by "guiding" them spontaneously. They take their hands and move through the problem solving events with them. In other words, they provide them with tactile-kinesthetic input which evolves during the event and transmits information about the changes in the surroundings necessary to solve the problem. As a result, the behavior of the children changes. What does this mean?

A big dog sleeps on the ground next to a newsstand. A little girl, 18 months old, sees the dog. Her father is busy buying a newspaper. She approaches the dog but stops at a safe distance to just look. It is obvious she would like to pet the dog, but at the same time, she is afraid to do so. Her father sees the situation, and realizes the problem. Spontaneously he takes his daughter's hands and strokes the dog once, twice, several times. Then he lets go of her hands. She hesitates for a moment and then pets the dog by herself. Her behavior has changed.

What has happened?

The little girl saw the dog – visual information! This information was not sufficient to answer the question: "Can I pet the dog"? Perhaps, it even caused a contrary thought: "I had better stay away from that dog". Now her father guides her hands and helps her touch the dog – tactile-kinesthetic information!
This sort of touching – tactile-kinesthetic input – can transmit all kinds of information about surfaces and their qualities, e. g., the softness of the fur, the warmth of the animal, and the "electric" tension derived from caressing the dog. There is also the movement of the animal – the regular up and down of its body when it breathes. In any case, the child seems to have received *enough tactile-kinesthetic information through guided petting that she is able to*

change her behavior. Her fear turns into confidence, and she pets the dog by herself.

Two fathers are in the woods with their sons, each 2 years old. They are playing hide-and-seek, but the boys are too small to play the game independently. What should the fathers do? They take their sons by their hands and guide them through the activities. One father hides behind a tree with his child; the other one leans against a tree with his, and they cover their eyes with their hands. Together they count to ten and then go to find the other two. Where are they hidden? After a while we can hear shouts of joy, for the boys have found each other.

The shouting and laughter and the obvious enjoyment of the two boys – all of this – allows us to assume that when being guided the boys got enough information to take an active part in playing the game. Both understood, even though they did not produce the playing themselves!

These examples describe *changes in behavior after receiving tactile-kinesthetic information* during a problem solving event. In the first example, the change of behavior was observable in the *production* of a movement – the girl petted the dog. In the second example, the change was seen in the *participation* – the boys played the game.

In addition to these two possibilities of observing changing behavior, there is a third one. I can observe the *facial expression* of someone who is being guided through an event. While a person is touching something, the change in facial expression can be interpreted as meaning that the person is *feeling* something. The following examples describe such observations and interpretations:

D., 5 years and developing normally, tries to peel a melon. His mother is close by.

He begins to cut the melon. His facial expression shows attention. He feels the resistance of the rind.

This is difficult! His mouth becomes tense. He feels that the knife is not cutting.

Now his mother guides his hands. He feels the knife go under the rind – feels the changes in resistance – and his face relaxes. ▶

E., 3 years, 6 months, is D.'s younger brother. He would like to cut the rind from a melon, also. His mother guides his hands.

◀
His eyes watch what happens – what his hands are doing – and looks at what he feels cutting the rind from the melon.

What he feels becomes familiar! He smiles.

As the knife goes around he must pay much attention. The tactile-kinesthetic information changes. . .

168

and changes again – the knife slices around the melon to the other side.

Here – is resistance! He feels the knife touch the plate. He looks to see what is happening.

Another piece of melon rind is cut. He looks at the changes produced by the cutting; he feels them.

◀

It seems that D. and E. are at their task with full attention. They feel; they experience the causes and effects. Often they take over and do parts by themselves. Sometimes they hesitate, the task becomes too difficult. Then those in the environment intervene – here it is mother – and help them by providing better tactile-kinesthetic information.

It seems that the children just described received important tactile-kinesthetic information when solving problems in daily life, e. g., petting the dog, playing hide-and-seek, and peeling a melon. *They received this information, either when they performed the movements themselves, or when they were guided in producing them.* In either instance, behavior has changed. We can conclude that the children had *learned*.

In the next paragraph, we will discuss the differences among several early levels of *learning* that can be observed when one receives tactile-kinesthetic information.

1.1.1 What I Feel Is Unfamiliar to Me

Let's begin with touching. The reader is reminded of the *unusual behavior* which those who are perceptually disordered may show when they touch something. One behavioral pattern is to "look away". At the same time, one can observe an increase of body tension (see Part II, paragraph 2.1.1 "They Withdraw from Touching, They Become Tense and Look Away).

It is concluded that on a *first level* of learning, children and adults can obviously receive an *overload of new tactile- kinesthetic information*. When this happens, tonus increases rapidly. As a result, there are quick and jerky movements, and the eyes look away most of the time.

Those with a severe disorder show this reaction characteristic of a first level of learning almost every time they touch something. Consequently, it is difficult to work with them.

1.1.2 I Feel and It Becomes Familiar; Now I Can Also Look at It

When I work with perceptually disordered children and adults by taking their hands and bodies to guide them through a problem solving task, we often touch material which is unfamiliar to them, e. g., something wet and slippery. This can elicit a first level reaction: Body tone increases rapidly and the eyes look away. As I continue guiding the problem solving task, body tone decreases. They relax. Their eyes begin to look at what is happening with the hands.

F., 23 years, is perceptually disordered and cerebral palsied. She is being guided. With her left hand she touches a piece of bread on the table.

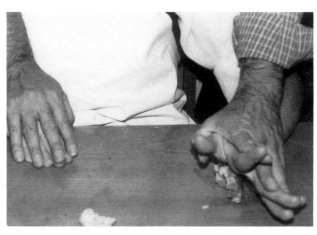

Her tone increases rapidly – her fingers spread out.

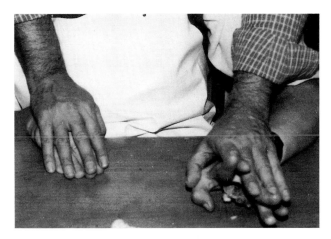

The touching continues – her body becomes more relaxed,

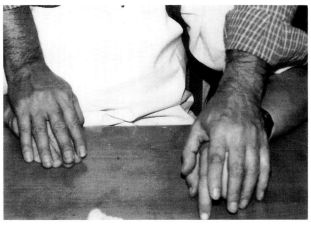

and her hand now closes around the bread we are touching.

This behavior characterizes the *second level* of learning through feeling. When touching continues within a problem solving event, it allows the person to become familiar with the tactile-kinesthetic information about that specific material. The amount of unfamiliar information decreases. There is only just so much new information that children or adults can see *and* feel at the same time.

We can, therefore, utilize the information about changes in body tone and direction of the eyes to judge whether those who are perceptually disordered are passing from level one to level two in learning. This point can be clarified further by discussing what is meant by changes in body tone and changes in the direction of the eyes.

Body tone changes. When we guide patients who have too much tone (hypertonic) in solving a problem of daily living, their movements become smoother during the guided task. If the patient is hypotonic, we can observe an increase of tone. In either case, body tone becomes more adequate or normal during guided problem solving events.

Such observations strengthen the statement: Within guided problem solving events, the body tone changes in the direction of normality. This only happens, though, if enough information is given about changes in resistance!

H., 7 years, is a severely cerebral palsied child who is in a special school. I intend to have him squeeze an orange to make juice. The physical therapist helps me position him on my lap. She shows me how to hold his legs firmly. She says that this is necessary because he gets very tense when touching anything. If he responds by jerking, I may not be able to control him.

In front of us are several items: oranges, a knife, a cutting board, an orange juicer, and a cup. I take his hands and begin the task. We grasp the orange and put it on the cutting board. We pick up the knife and cut the orange in half. Putting one half of the orange on the juicer, we press it down and make juice. We pour the juice into a cup and drink it. In between, we explore the cut orange with his fingers, hands, and mouth in the same conventional and unconventional ways that a young child would. Whenever we begin a new step, e.g., when he touches the orange or the knife, he becomes very tense. However, this decreases as soon as we do something with the object that is directed toward our goal. The longer we work, the more his tenseness decreases and the more relaxed he gets. Finally, I can let go of his legs.

The interpretation of his reactions: When touching something for the first time during the task, the tactile stimuli were unknown to H. The same interpretation applies to his jerking response to something or someone touching him in a more free situation. However, when something meaningful was done with him and the objects or materials were touched, he became more familiar with that tactile stimuli. Therefore, it is assumed that the high tension was elicited by his lack of familiarity with what we were touching. Tension decreased as he became more familiar with it.

Ch. is a 23-year-old patient with a head injury. He is sitting in his wheelchair; his trunk and head are slouched way down low. The physical therapist complains that it is so difficult to stimulate his head control.

Since Ch. is thirsty, we take him to the vending machine where he can buy something to drink. I guide his hands to his wallet and help him grasp the coins he needs. Where should he put them? I guide his hands upward with the coins. (The slot is way up high on the machine.) His trunk stretches and his head lifts up – a little, then more and more – but it still isn't high enough. We haven't yet reached the slot. We stretch some more and now he can push in the coins. It is the same with the selection button. He has to reach way up high to choose the drink he wants.

The physical therapist is astonished. Ch. shows much better body tone and such beautiful lifting of his trunk and head!

Closely related to body tone is *motor performance*. It is impressive to observe how many movements brain damaged patients can make during problem solving tasks! Here is another example:

Mr. F. is severely impaired in motor performances after suffering a head trauma. He is being guided. The task: A bell which will be used to call the students from recreation should be hung on the wall.

He follows the procedure with full attention. While still being guided, the moment comes for him to hang the bell on the nail. He is sitting in his wheelchair. He lifts the bell up with his arm. The nail is farther up. He stretches more – stretches up and up. His trunk is stretching; his arm is stretching; his fingers are stretching. Finally the bell is hung!

During a conference the next day, Mr. F.'s physical therapist watched a video taken during the task. When she saw how he stretched his body, including his arms and fingers, she was overwhelmed and expressed astonishment, "I've practiced a hundred times with Mr. F., and he was never able to stretch like that"!

With tactile-kinesthetic input during problem solving tasks, motor performances get better. Such improvements are not restricted, though, to just the functions utilized during the guided event. There is transfer. This is an astonishing phenomenon! After a guided event, patients improve in walking – even when they have not practiced walking during the event.

N., 23 years, has had a severe head trauma. When guided, she can walk a few steps with difficulty. The physical therapist brings her to occupational therapy. He walks with her from her wheelchair at the door to the table in the middle of the room. There she sits down and stays in that position during the guided task. At the end of the occupational therapy, the physical therapist comes to help her again. He walks with her from the table to the door where her wheelchair is waiting.

Her walking has improved; the therapist expresses his surprise at this.

As mentioned earlier, another behavioral change, which belongs to the second level of learning, is the *direction of the eyes*. When children and adults are guided in a problem solving event, the hand is often moved toward an object to be grasped. The object is needed to perform a step within the event. In this case, the eyes follow the hand with a short delay. With each repetition, the time of delay usually shortens until the eye follows the hand immediately.

A., 9 years and perceptually disordered, is coring an apple. She is being guided through the task.

She has just cut a piece out of the apple. Her eyes are still looking at the spot on the cutting board.
Her hand is already near the bowl.

Now her eyes follow her hand.
She looks at what her hand is doing at the bowl.

Observed changes in the direction of the eyes when the body or hands are involved in solving problems can be attributed to *underlying perceptual processes* as analyzed in the following:

In the first situation, the hand was touching but the eyes looked away. Whatever is touched is interpreted as being unfamiliar to the perceiver – a new stimulus. Consequently, it is explored exclusively through the tactile-kinesthetic system.

In the second situation, the hand was touching something, and the gaze was directed toward what was happening with the hand. The tactile-kinesthetic input is interpreted as being more familiar. Thus, it becomes possible for the perceiver to coordinate visual with tactile-kinesthetic information. In addition to the modality specific information, intermodal information (inter = between; modal = sensory) is received.

The next level of learning is characterized by a *long duration* of feeling *and* seeing when problems are solved.

1.1.3 I Feel and Look – I Look and Feel

As was previously mentioned, at this *third level* of learning, we can observe a strong *relationship between the hand and the eyes.* The eyes watch what the hand is doing, almost continually. When the hand grasps a new object, the eyes rapidly change their direction to watch the event.

They can focus a long time on what the hands are doing. This obviously means that perceptual activity can now utilize a *time dimension.*

The lasting coordination between eyes and hand can also be interpreted as an expression of a *sequential organization of intermodal information.* Problem solving demands a sequence of activities which has already been described in Part II, paragraph, 2.2.3 "The Sequence – When Something Is Missing or When You Cannot Go Back"? If a lengthy fixation of the eyes on the hand is observed, then we can surmise that the person is coordinating the tactile-kinesthetic information with the visual information from the same event. This means that, in addition to the modality-specific tactile-kinesthetic information, he or she can now integrate intermodal information of a visual-tactile-kinesthetic kind. Furthermore, this intermodal information will be ordered according to the sequence of actions of the event. This is called *serial organization.* Thus, this third level of learning is characterized by important intermodal and serial performances.

Once again we are reminded of the importance of the ideal learning situation provided by problem solving events. Examples in Parts I and II of the book illustrated that several sensory modalities receive information when problems in daily life are solved. Recall the variety of situations involved in preparing meals where one is feeling, smelling, tasting, and also seeing and hearing. All of the input occurs simultaneously. In addition, there is always the time component: I hang up the clothes I washed earlier, and I will take them off the line later on when they have dried.

All of the perceptual processes underlying such events, along with the information they provide, become meaningfully interrelated within the event and the corresponding problem solving activities.

A., 9 years and perceptually disordered, prepares a filling for a lemon. She is being guided.

She pours yogurt into a bowl.
She is touching/feeling and watching!
The filling is ready.
She presses the filling into the lemon. She touches/feels. . .and looks at what her hands are doing throughout the long event.
Visual information becomes interrelated with tactile-kinesthetic information and both of them with the corresponding causes and effects of the event.

The more children and adults experience "touching/feeling and looking" within guided problem solving events, the more often will they show behavior which is characteristic of the next higher level of learning.

1.1.4 I Recognize What I Feel and Continue the Movements; I Anticipate Them

A new kind of behavior becomes observable and can be interpreted as *recognition* of something felt. While touching something during a task, both children and adults who are perceptually disordered may smile or offer resistance by turning their heads away or pushing the object away.

175

A., who was coring an apple and preparing a filling in the last two examples, is performing another task.

Some nuts are in a bowl.
She mixes them with the yogurt.
"Oh! This is sticky"!
She tries to remove the sticky material from her fingers. She begins to smile. We interpret the smile as recognition of the sticky substance.

Another behavioral characteristic of the fourth level is the *spontaneous continuation of a movement begun with guidance.*

A., from the preceding examples continue to make her filling.

Now she takes some nuts out of a narrow jar. Her fingers meet strong resistance from the sides – this is difficult! She is touching/feeling and looking.

There! Now she can continue the movements by herself!

Sometimes, those who are perceptually disordered like A. can *take over and continue* a movement which began with guidance. When guiding someone through a problem solving event, one can feel when this moment arrives. The body tone of the guided person increases, and the arm begins to move in the direction of the task. If this happens, one should let go of the person's hands but stay close by in order to take over the guiding again as soon as it may be needed. A.'s teacher used this method, as shown in the photograph.

This fourth level of learning is also characterized by the *anticipation of an action belonging to the actual event*. Anticipation presupposes recognition of the movement when felt. It is more difficult, though, since it includes a *more complex time dimension*.

The soup is steaming. The child looks at her father knowing and anticipating that he will blow into the soup to cool it so she can eat it. She is waiting *for an action from someone in her environment. She is waiting for something that is not present at the moment – for something that will happen.*

Waiting for something that will happen is called *anticipation* (Piaget, 1958; Piaget & Morf, 1958). Recognition, on the other hand, refers to a happening *at the moment*. I am feeling something sticky. I know that I am now touching something. Anticipation goes *beyond* recognition since it includes something in the *future*. However, it isn't required that the children or adults be able to produce the anticipated action.

The kind of anticipation which is characteristic of performance at this level of learning is very simple. It refers to an action which will *immediately* follow an actual action.

The soup is steaming. In the next moment, father will blow on it.

The following example shows that anticipation precedes production for normal children, as well.

J., 2 years, is having her birthday celebration.

The candles are lit on her cake, and she understands what will happen to them. She recognizes the situation and anticipates that the person next to her will blow out the candles, for she cannot do it herself.

In another situation a peach is divided in half, and now each part should be cut into smaller pieces.

M., 2 years, 4 months, touches the peach with a knife. She understands the situation, recognizes it, and expects – anticipates – that someone will cut it, but does not do it herself. This is still too difficult for her!

1.2 I Touch and Allow for Touching

The four levels of learning which have been described emphasize the possibility that children and adults can change their behavior as a result of tactile-kinesthetic information. We can interpret such behavioral changes in children as becoming familiar with the actual situation and the corresponding event, and in adults, as becoming reacquainted again.

In the following paragraph we will reflect on the problem of how to guide the body of a person in order to transmit more adequate tactile-kinesthetic information.

1.2.1 I Guide from Behind

Guiding from behind *facilitates* the guiding. In this position I can transmit and receive maximum tactile-kinesthetic information.

I *transmit tactile-kinesthetic information* if I embrace the body when I guide. Consequently, the surface for transmitting tactile-kinesthetic information is broader and has a *positive* effect. When I direct all my attention toward the person being guided and toward the event in which we are involved, I transmit my directed attention to him or her. When I am very calm, the one being guided will feel my calmness.

But there can also be a *negative* effect. When I am very tense, I transmit my tension to him or her and he or she will become tense, too. I will also cause a negative effect when my attention is not focused on the individual and event but rather on something else, e. g., when I am waiting for a telephone call or when I am talking with another person in the room. In such cases, the person being guided will resist.

To be calm internally and to direct full attention to the one I am guiding can be difficult to achieve. For example, a mother can-

I put my right hand on the right hand of the child and my left hand on the left one. I concentrate on the tactile-kinesthetic information I am receiving from the child's body and through it from the event. I do not need to see the child's face – the body tone indicates the amount of attention being paid to the event.

Even when a child stands, I guide from behind.

not guide her child if she is continually under stress. Her stress could be due to her other children asking for her attention or to additional factors that limit her capacity to process further information required for guiding. It can be a similar situation, too, when teachers have to care for a group of pupils and must divide their attention among them.

During *problem solving activities it is essential that children or adults receive tactile-kinesthetic information* from guiding, not only for attending to the cognitive part of the event, but also for the emotional content of those situations. Due to this information the person has a *holistic cognitive and affective experience*. Likewise, parents who report their experience in such situations, say they feel very close to their child.

Consequently, I am not only transmitting, but – as was mentioned in the beginning – *am also receiving tactile- kinesthetic information*. I am feeling the reactions, the attention, the changes of tenseness, the warmth, the calm or the unrest of the person I am guiding. There is an exchange of tactile-kinesthetic information going on between the person being guided and the person doing the guiding. This is a *tactile-kinesthetic interaction*.

1.2.2 They Have Two Hands, a Mouth, and a Body

One can guide any part of a person's body: trunk, legs, feet, arms, hands, and fingers. The decision as to *where* one should guide at any given moment depends on the person being guided and the activities of the event being performed at that moment. Most frequently I guide the hands – my right hand on the right hand and my left hand on the left hand.

They Have Two Hands, ...

When I manipulate something, I usually utilize both of my hands. The right hand performs a different part of the activity than the left one. I should proceed in a similar manner when I am guiding: To guide someone in eating, I may place my left hand on his or her left hand to hold the plate; I put my right hand on his or her right hand to put the food on the spoon or fork. To guide someone in hammering a nail, I place my left hand on his or her left hand to hold the nail; I put my right hand on his or her right hand to manipulate the hammer. To guide someone to cut a slice of bread, I put my left hand on his or her left hand to hold the bread; I put my right hand on his or her right hand to manipulate the knife.

At all times, I am aware that a human being has two hands. The information of each hand goes to a different hemisphere of the brain. It is important that both hemispheres learn how to *work together*.

In this photograph the guiding is being done incorrectly. *Only the left hand of the person is being guided. The other hand is passively on the table. The bowl is being held only by the person who is doing the guiding. Thus, important information is being left out; important information is not being transmitted.*

Here the guiding is done correctly. Both hands are participating in the event. Information concerning the different steps is transmitted to the brain from both hands – the whole brain is collaborating.

Try it for yourself. Have someone guide you – once incorrectly and once correctly. Close your eyes when trying this. Can you feel the difference?

The goal of this kind of work – to emphasize once more – is to transmit tactile-kinesthetic information. This involves *perception. We feel equally well on both our right and left sides. Of importance here, though, is the cooperation between the right and the left sides.* With this in mind, we must clearly differentiate between perceptual performance and motor performance. In motor performance we are dealing with a dominant side, but in tactile-kinesthetic perception we are not.

When I guide both hands, I have to be careful that I put *my fingers exactly on the corresponding fingers of the person I am guiding.* In this way, his or her fingers feel the surroundings as I feel them and embrace an object as I embrace it. I have to constantly check to ensure enough time for the positioning my fingers. Two pictures illustrate this point:

This is how it should not be done. The left hand of the person being guided is held across the wrist as if it were a tool. Wrong!

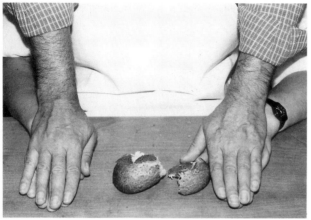

This is how it should be done. The hands of the person doing the guiding rest on the hands of the person being guided, forefinger on forefinger, little finger on little finger, and thumb on thumb. Bravo!

. . . a Mouth, . . .

Just like normal children, perceptually disordered children use their mouths to go through a stage of spontaneous exploration of the environment. The duration of this kind of exploration is often very long for perceptually disordered children; it can last for years. In normal development, it generally covers only the first two years. Perceptually disordered children usually begin this stage at an older age. They still use their mouths at an age when normal children do not. This behavior is often embarrassing to those around them and is therefore frequently misjudged as being a bad habit. Often, attempts are made to wean them from exploring with their mouths.

However, *the utilization of the mouth is very important for development.* Instead of weaning those who are perceptually disordered from using their mouths, we should stimulate them. The mouth is not only important for exploring the qualities of objects, but it can be used for manipulation.

Eating is a favorable occasion for exploring *qualities* with the mouth. Such an exploration can be included when guiding a perceptually disordered child, or even an adult, in preparing meals.

F., 23 years, perceptually disordered and cerebral palsied, is being guided during the preparation of a cheese sandwich.

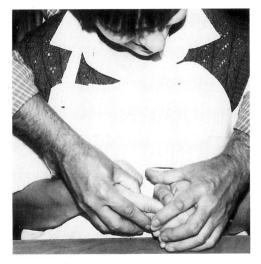
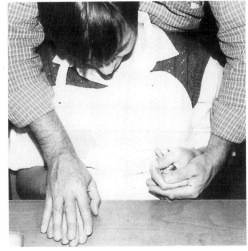
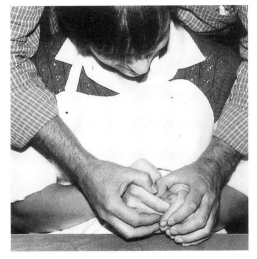
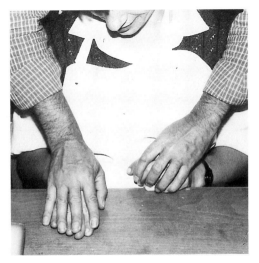

She presses the piece of cheese she has broken off into the bread.

The right hand explores the stable support of the table. The left one holds the bread and turns it over into the eating position. At the same time she feels the changes in resistance between the table and her own body.

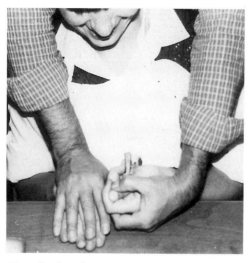 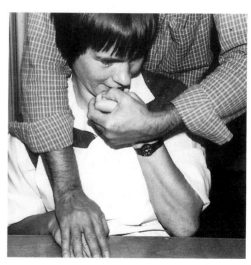

Now the hand moves along the body to the mouth.

"Mmm! This tastes good"!

When being guided during meal preparation, some perceptually disordered children seem to push food into their mouths in a greedy manner. These children are often poor eaters. Their behavior looks almost compulsive. Our experience, however, shows that this is not the case. When these children have had enough experiences with the qualities of their surroundings, especially through many explorations with their mouths, the frequency of this kind of greedy behavior decreases spontaneously. At the same time, we observe an improvement in their eating habits. The important thing is that we help them to experience a variety of movements when being guided. In this manner we can vary the tactile-kinesthetic information they receive through exploration with their mouths.

Acting upon is another use of the mouth that can be observed at any age level, even among adults. The mouth can help whenever the hands are not sufficient. For instance, I can use my mouth when I need some force, such as when I want to crack a peanut. In this situation, the mouth is not utilized for exploring the qualities of my surroundings, but rather as a kind of tool or resource for performing an activity.

Hypotonic children often use their mouths for manipulation when their hands cannot produce enough strength. This can happen with such frequency, that those around them incorrectly call it compulsion.

T., 9 years and hypotonic, is being guided as he explores the contents of a small box. Another box is inside it, and it contains a small tin box. With the box in his hands, T. repeatedly holds it up to his mouth and bites on it. I guide his fingers so that with the help of his mouth and his biting the box can be opened and the contents taken out. T. pays full attention to this task.

W., 9 years and perceptually disordered, tries to open a bottle by pulling out the cork.

He does not succeed. The cork does not move.
He tries with his mouth.

Again no success! What can he do? W. feels and looks.

He tries another time; he bites on the cork.

Without interrupting the event, the teacher takes both of his hands to perform a small variation of the movements of his mouth and the object.

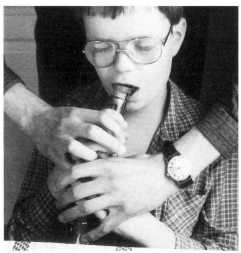

And now he succeeds. The resistance of the cork yields. ▶

A., 9 years and perceptually disordered, is supposed to squeeze a lemon. She is being guided during the task. With both hands, she tries to press the lemon hard enough so that the juice will come out.

Her hands are too weak. Why not having her use her mouth? The hands holding the lemon are guided across her body to her mouth.
▼

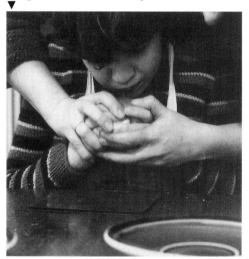

▲
She bites into the lemon. "Yuk! This is sour"!

◀
But now her teeth hold something tightly.

In this way she has not only performed an action with her mouth – opening the lemon – but has also experienced something about the quality of the lemon.

Similar events happen in the following example of a normal child:

D., 5 years, is being guided by his mother in preparing a melon.

Together they touch and press the rind of the melon. D. smiles in anticipation.
The wet and slippery part of the fruit that he sees and feels can be grasped and taken off.
"Maybe my mouth can help, too". ▶

He touches it with his mouth. It smells; it is wet and slippery; it can be bitten, sucked, and taken off. He can eat it.

The rind resists; it is the stronger part – the part that doesn't yield, the rough part!
▼

...and a Body

Like the hands and the mouth, the whole body serves to transmit and receive tactile-kinesthetic information. Whenever the whole body can be used to solve a problem, we should use it. The more frequently we vary body positions, the more information we will receive. But here also, as has been emphasized throughout this book, we have to be conscious that each movement of the body must be embedded in a meaningful event.

In the following example, A. is supposed to sweep a dirty floor with a vacuum cleaner. In order to put the parts together, she has to take on different body positions.

A., 10 years and perceptually disordered, is being guided.

◀
"What a long thing that tube is"!
She kneels on the support to put the parts of the tube into each other,

▼
and now she can begin to sweep.
She stands up and bends over.

"There, under the table, it is also dirty"! She stoops down and kneels.

I will change positions as often as possible, depending on the person being guided. I will sit on a chair, stand at a table, sit on the floor, or walk. I will reach for an object across the table, bend to pick something up from the floor, kneel or lie on the floor to clean a low shelf, etc. I will get something we need from the cupboard and stretch my body to reach it. I will put something into the garbage pail, open the door with the person being guided when both hands are full, or climb on a chair to get something that is higher up. All of these positions have to be *meaningful* parts of the event. All of the different activities – walking, kneeling, climbing – require that I guide the trunk, the hips, or the legs. To meet that requirement I will often ask: How would a normal, young child climb on the chair, kneel on the floor, or bend to pick something up? To answer this question I must *observe small children over and over again*.

To Summarize:

By using examples of both normal and perceptually disordered children, we *note*:

- Tactile-kinesthetic information received during solving problems in daily living activities allows important changes in behavior which can be interpreted as learning.
- Such learning happens in levels that can be observed in persons who are normal and in those who are perceptually disordered.

We *interpret*:

- Children and adults who are perceptually disordered can use tactile-kinesthetic information within problem solving events in daily life.
- However, when they are left on their own, they receive an insufficient amount of tactile-kinesthetic information (see Part II, "Failing in a Wirklichkeit").
- Consequently, they depend on us to transmit such information to them.

- Within daily life events the transmission of tactile-kinesthetic information can be done by guiding different parts of the body.

We *ask* ourselves: What kind of information do we take out of problem solving events in daily living situations?
In the next chapter we will think about this aspect.

2 I Feel and Act Upon

2.1 I Feel the Wirklichkeit

In order to learn from tactile-kinesthetic experiences, they have to be *meaningful*. Experiences become meaningful when they relate to solving the problems in daily living. There are always some problems. I want to open the window – it sticks. I take an egg out of the box to boil it – it has a crack. I am sewing – the needle drops. To solve these problems I have to become familiar with the actual situation. What kind of window is it? Do I have other eggs? Where has my needle fallen? I also have to know about *acting upon*. How do I open that window? What movements do I have to perform so I can get it open? I have to feel all of this – the surroundings in the actual situation and the acting upon. In summary: I have to *feel the Wirklichkeit.*

2.1.1 The Wirklichkeit Includes the Surroundings,...

To solve daily problems we have to change the Wirklichkeit. To elicit these changes we have to adapt our actions to the actual situation. This requires *continuous tactile- kinesthetic interaction with our surroundings.*
When I guide children or adults in solving problems in daily living, I try to pick up tactile-kinesthetic information about the qualities of the object we have to handle. We

need this information in order to adapt our movements – touching, embracing, moving – to the material, the shape, and the changes we want to elicit. Such tactile-kinesthetic exploration is, therefore, a prerequisite for adequate acting upon.

The following example illustrates the strong interrelationship between the tactile-kinesthetic exploration of the surroundings and acting upon:

D., 11 years and perceptually disordered, is stuffing tuna fish into tomatoes (see Part II, paragraph 2.1.1 "They Withdraw from Touching, Become Tense and Look Away").

▶

When he opens the can, he feels changes in resistance. He looks at his hands. The information he is receiving includes something he is familiar with. He holds the can – it provides stable sides and a support.

The can is open. He stabs the fork into the can. It gives him familiar changes in resistance. He feels and looks.

He briefly touches the contents and changes his facial expression to show a slight increase of tension. "This is wet and slippery"!

"And now! What is this"? A tomato! He knows the tomato from having seen it before – a beautiful red – but to touch it? In addition, the tuna slides off the fork! ▶

"No"! This is unfamiliar... ▼
...and there is hardly any resistance to feel. His hands are in the air; the fork is in the air!
"Where are my surroundings?
Where is the stability"?

He stabs the fork into the can again. This elicits changes in resistance.
He closes his eyes. His whole attention is directed toward what he feels – toward acting upon.

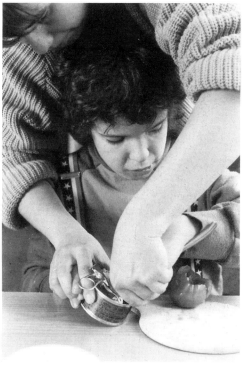

2.1.2 ...and Changes in Resistance Are Needed

The one I am guiding and I *see* a table with "colored spots" on it. Can we *touch* the spots we see? We touch the table first. Is it a stable support? We touch the visible spots on the support. Can we embrace these visible spots? We try. We are successful. Together we *embrace* one of them, *displace* the "thing" on the support, *take it off* the support, *and move* the "object". Each one of these steps causes *changes in resistance*. Together we feel them – *perceive them*.

We described Mrs. R. in an earlier part of the book concerning failures (see Part II, paragraph 2.1.2 "They Know About the Rules of the Stable Support and the Side"). Mrs. R. talked almost continually. The therapist guided her in taking the brush, moving it through the air to the pan holding the colors of paint, and then through the air and onto the piece of paper. Mrs. R. became silent each time she elicited a maximum change in resistance – this happened even when she was being guided. Try the event with the brush. Close your eyes so you can concentrate entirely on the tactile-kinesthetic input. Have someone guide your hand, embrace the brush, move it through the air, put it into the paint, move it through the air, and put it on a piece of paper. Now repeat the steps. Even though you are being guided, you will feel all the changes in resistance each time.

The examples which follow describe some situations with changes in resistance that we feel within problem solving events in daily life. In the first example, the changes will be elicited by the *hands*.

A., 9 years and perceptually disordered, is being guided in removing the paper from a lemon.

She has taken the wrapped lemon. With her left hand, she holds it on the stable support. The right hand approaches,

and touches it. She feels the paper around the lemon, and at the same instant, she continues to feel the stable support. She embraces the wrapping until she can feel its resistance; then she takes it off.

Unwrapping the lemon required touching, embracing, and exploring the qualities of the lemon and its wrapper. The tactile-kinesthetic information received by these actions included both familiar and unfamiliar features. With each tactile-kinesthetic experience, the surroundings are perceived and analyzed. They become more familiar – more intensive, more extended.

Causes and effects: The wrapping hides the lemon. The wrapping and the lemon belong together. Then comes the separation – the change! The soft and yielding part is taken off – a change in resistance – and is crumpled until the total resistance is felt. Then it is released on the support of the table. Through such an insignificant event in daily life, A. is *acting upon her surroundings.* In this manner, the surroundings become more and more of a Wirklichkeit to her – a Wirklichkeit which is moving and changing again and again.

By being guided through even such short events, perceptually disordered children and adults can master some of the needs in their daily living. In this sense, such experiences are a help for them to become better prepared for dealing with their lives and, at the same time, are a base for progress.

In the next example, *the hands and mouth* are used to elicit and to perceive changes in resistance simultaneously:

E., 3 years, 6 months and a normal child, is making melon balls as part of the preparations for dinner. His mother guides him when he is unable to perform the activity by himself.

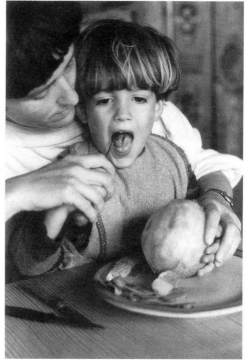

He cuts into the melon with a melon ball scooper; there is resistance, yielding of the resistance, and resistance again. "How fascinating"!

"Is there anything on it"?
He elicits and feels the changes in resistance with his mouth and his tongue!

 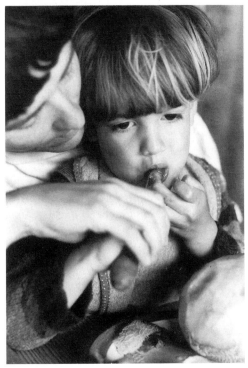

▲
"Mmmm"!
He closes his eyes and savors what he feels and tastes.

Again he feels the changes in resistance elicited by his fingers, his tongue, the melon ball scooper, and its contents. His eyes look nowhere else. The quality of the melon provides so much to feel and *there is so much involvement with acting upon it.*

▶

With his fingers, he presses harder on the melon in the scoop – he feels it. "Will the resistance change again? Maybe I can put this tasty thing into my mouth"? He closes his eyes.

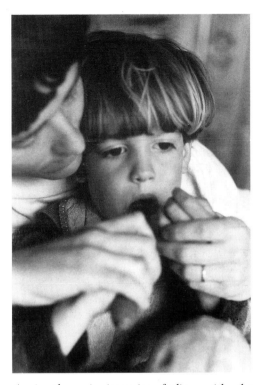 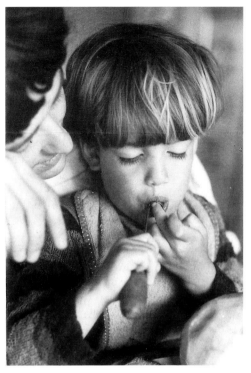

Again there is intensive feeling with the tongue – resistances and qualities! The eyes are open; they look into a void.

"Sucking might help"! He embraces the scoop with his lips, and at the same moment, he presses and feels with his fingers. His attention is entirely directed toward what he feels. He closes his eyes again and doesn't even notice that his mother is no longer guiding him.

Whenever I am guiding perceptually disordered children and adults through problem solving events, I consider the changes in resistance between our bodies, the support, and the object. I am conscious that I have to embrace an object before I can remove it. Taking off can only occur if I know the qualities of the support and the sides. Therefore, I always explore the stability of the support and the sides before I start to act upon an object.

2.2 They Know About the Rules...

In Part II, Chapter 2 "We Observe: They Have It and Yet They Don't Have It," we referred to observations where perceptually disordered children and adults knew about the rules of touching and of acting upon. We concluded that they are competent with the rules, but that they often do not reach the level of performance due to a lack of information (see Part II, paragraph 3.3.4 "...and the Competence Does Not Become Performance"?). This is an important statement to remember when working with perceptually disordered children and adults.

When we succeed in providing them with better tactile-kinesthetic information, they should be able to apply these rules.

2.2.1 ... of Touching...

When I guide someone I have to consider what kind of information would be needed if I could not rely on visual information but had only tactile-kinesthetic information?

Before I begin a movement, I explore the stability of the support. "Where is my body? Am I standing or sitting on something that is stable"? I move my body, my feet, and my hips to touch the support! When I guide, I explore the support with the person I am guiding in the same manner as for myself. For instance, I move a little on the chair. "Is the chair shaky? No, it is firm". Now I put my hands on the table and feel its resistance. "Is it stable"?

When I am sure about the stability of the support, I begin to consider the *problem* and to solve it.

I move. In addition to the support, the *sides* now become important. When I try to provide a perceptually disordered person with adequate tactile-kinesthetic information, I have to pay attention to an important *principle*: Whenever I change the position of my body, I need another *stable resistance as a reference surface for my body movement* – a resistance which is in addition to the support I am standing on.

When I am *sitting* on a chair at a table, I will use

- *my hands*: I put both hands on the table as my stable support and begin by moving one hand toward the object. During this movement the other hand stays on the support to assure its stability. After one movement, I change hands.
- *my body*: I press my body against the edge of the table to assure myself of its stability. In this position my hands become free to move.

When I am *standing*, for instance, while doing the dishes or while cleaning the refrigerator or a shelf, and want to move through the room, I use

- *my hands*: This is similar to the sitting position. One hand is always assuring the stability of the side, for instance, the wall; the other one searches for the next stable position. Then I change them.
- *my body*: I press my hips, for instance, against a stable side; I can now move my hands freely.

This principle comes from the knowledge that *perception is relative* (Piaget, 1961). Changes can only be perceived in relation to something that is stable. This is not only valid for the tactile-kinesthetic sensory system but also for other sensory systems. When I am sitting in a train and look out the window, I see houses, trees, and people. I know that they are not moving, but when the houses and the trees and the people disappear, I know my train is moving. If another train is passing by my train, though, I do not know if it is my train or the other one that is moving.

We have been reflecting on how normal children learn to search for objects. During the months before they can walk, they acquire a multitude of touching experiences with objects which have been placed on the same support with them. In this way, the *support with its stable resistance* becomes an important *reference surface* for the coordination of both the right and left hands and all ten fingers, for providing important information about qualities of the surroundings, and for the possibilities of acting upon them.

When the support does not become embedded in the information of touching, the various and obvious kinds of behavior – the deviancies – will develop (see Part II, paragraph 3.2.4 "The World Does Not Become a Surrounding World"). Instead of the unity and the harmony of finger movements, there will be confusion. The use of one hand will not develop into the use of two hands, which is a prerequisite for the development of the "unity of two hands" (see Part I, paragraph 1.4.3 "One Hand - Then Two - And Yet a Unity"). Instead of embracing, children seize objects with just two fingers as if they were a pair of tweezers, an expression of a deviant development, followed by increasing hectic behavior.

2.2.2 ... and Acting Upon

The surroundings change continually in daily situations. People move. I move. Objects are taken off the support, put back on the support at other places, and released. Relationships of neighboring things are changeable.

In Part I of the book, we described, step by step, how normal children get to know about taking off and changing neighboring relationships between things and persons through tactile-kinesthetic information. But this is not the case for those who are perceptually disordered! Their failures in everyday life include the difficulty in grasping the complexity of changes in neighboring relationships (see Part II, paragraph 2.2.2 "And Where Is the Neighborhood"?).

Consequently, when I take my time to guide perceptually disordered children or adults through a problem solving event, I must include the *taking off and exploring of the neighboring relationships in the guiding.*

Two questions arise in connection with this principle:

- How do I reach an object I need but which I am not touching at the moment?
- When I have reached my goal and do not need the object anymore, where do I put it on the support, and how do I release it there?

The first question: *How do I reach an object across a certain distance?* In Part I A, paragraph 1.4.4 "The Multitudinous Ways of Touching and Releasing," we described the behavior of little T. Through a variety of actions, she learned that an object she is not directly touching *stays in a neighboring relationship to her body across the support.* This means that *she can reach it over the support being shared by the object and herself.*

This principle has to be followed when I try to provide a person with better tactile-kinesthetic information when guiding him or her to solve problems.

F., 23 years and both perceptually disordered and cerebral palsied, has finished cutting some cucumbers. Now the pieces should be put into a pan. In order to do that, she has to get the pan. She is guided in solving that problem.

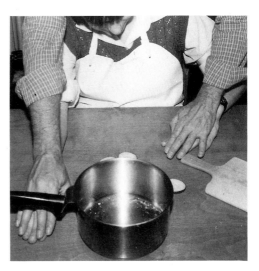

With the left hand she feels the stability of the support of the table while the right hand searches for side resistance on that support.

First the right hand, then the left one, finds the side resistance. It yields – relative to the stable support.

Both hands feel the resistance at the sides and, simultaneously, the stability of the support.

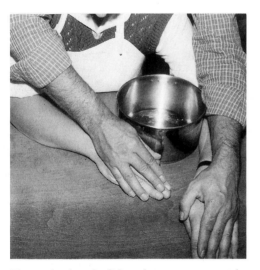 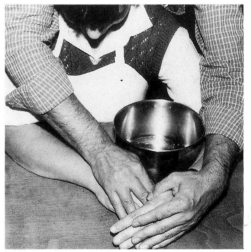

The right hand slides the pan across the stable support and the left arm gives some side resistance!

She moves it over the support in the direction of the trunk of her body!
The left hand takes over the movement.

Now the pan touches the body; the support provided by the table is still included in the performance – a stable reference surface!

The pan is brought closer to the body. It is in a niche between her arm and body. There are side resistances all around!
Now she can risk leaving the support and moving the pan along the edge and below the table in preparation for depositing the cucumber pieces in the pan.

The second question concerns *putting together and releasing*. When we are active in our daily lives in the sense we have described in this book, we are constantly taking off, bringing together, and releasing things. What happens in these actions? What kind of information do we use in order to perform them? Recall the development of normal children (see Part I A, paragraph 2.1.2 "I Separate and I Bring Together"). Children reach objects across a support. They explore how to take them off the support, how to separate them from the support, how to bring them together again on the support, and how to release them there.

The principle of releasing has to be followed when guiding perceptually disordered children and adults. *I do not release the object in the air but wait until I feel the support.* Experiment with this principle: Allow yourself to be guided in releasing an object you have in your hand. Guided, you explore the stability of the support with the hand holding the object. Feeling the resistance of the support, your fingers open almost automatically and thus, let the object go – release it! This principle is valid for every situation in which I have to release something in order to bring it together with something else. This is the case in the examples of the previous paragraph. Pieces of cucumber are put into a pan – the pieces have to be brought together with the bottom of the pan. This requires a release of the pieces as they touch the bottom of the pan – the support. The situation may be described as follows:

F., from the previous example, has grasped the pan. Now, the pieces of cucumber should be put into the pan.

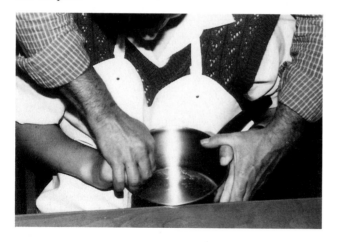

The pan is taken downward along and off the edge of the table,

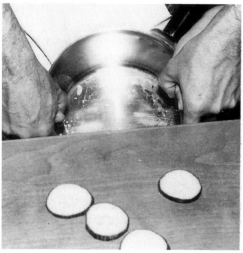
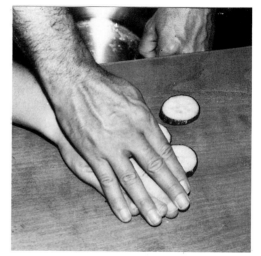

and then it is held firmly on her lap which is used as a support, and her body and the edge of the table are used as stable sides. The left hand assures this position...

for the next step so that the right hand can move freely on the table – the support.
She slides the pieces of cucumber across the table top to the edge where the pan touches it.

The resistance of the support changes as the right hand with the pieces of cucumber touches the inside of the pan.

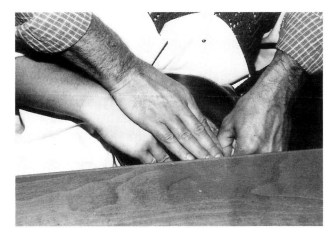

The hand with the pieces of cucumber moves down along this side resistance.
Alternatively in between this step, the left hand assures the stability of the position.

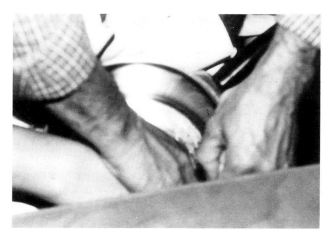

After the right hand reaches the bottom of the pan, it presses the pieces against the bottom!

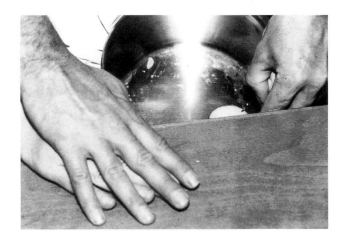

And here, on the newly felt resistance of a support, the cucumbers are released.

Releasing is often done with a *tool* which has been used but is no longer needed. This is illustrated by the next example. An egg- plant has to be cut, and then the knife has to be put away.

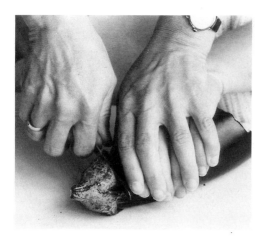

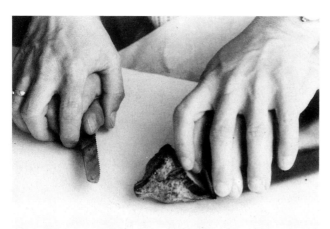

Before cutting, one has to feel the different resistances offered by the situation – on the support and at the sides. Then the cutting can begin.

The right hand, which is holding the knife, touches the support and feels its resistance.

It moves slightly backwards and opens the fingers to release *the knife.*

Another aspect concerning the discussion of neighboring relationships is that perceptually disordered children are often referred to us only after they have already become quite *hectic*. We remind you, the reader, of previous comments (see Part II, paragraph 2.1.4 "...and Don't Succeed In Embracing Things," and 2.2.2 "And Where Is the Neighborhood"?). These children do not include the support in their learning about neighboring relationships. *Instead of constructing tactile-kinesthetic neighboring relationships, they construct visual relationships.* This creates difficulties, e. g., judging distances in space. For instance, when I guide such children over a support and they cannot use this information, they react accordingly. The information is unknown and they become *panicky*. It appears in this situation that they function on the first level of learning (see Part III B, 1.1.1 "What I Feel Is Unfamiliar to Me").

Educators who have to work with these kinds of children need intensive professional help because they have to first *reverse the deviancy in development*. This requires beginning with *tactile-kinesthetically perceived neighboring relationships* - relationships of a *direct* kind, which include the body, a support, and an object. Such children cannot be guided along a support to an object by approaching the object and reaching for it over the support. These children need to get the object on the support as quickly as possible so that they can feel the object, the support, and their own bodies *simultaneously*. In this way they can slowly change their visually oriented neighboring relationship into a more basic *tactile-kinesthetic* relationship.

2.3 And Now I Can Change the Surroundings

Daily living brings problems and I have to solve them. Solving the problems demands changes. The changes have to be elicited – they have to be caused. For instance, if I stand in front of a closed door, I have to do something to open the door. I move my body and hands toward the doorknob. I have to turn the knob - the cause to open the door. The doorknob moves - the effect - because my hand turns it, and thus the mechanism unlatches the door and it opens.

The following example illustrates some cause-effect actions when utilizing the vacuum cleaner:

A., 9 years and perceptually disordered, cleans a room with the vacuum cleaner. She is being guided.

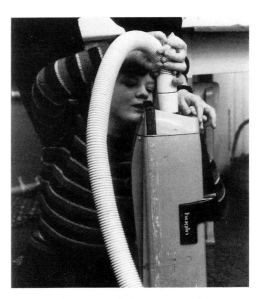

A. puts the parts of the vacuum cleaner together. She puts the tube into the opening until the moving tube stops and is firmly connected to the vacuum cleaner.

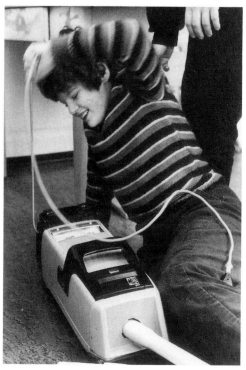

Now the cord. This action she can perform herself; she takes over the movements.

The cord is plugged into the socket – cause. There is no resistance; then there is total resistance – effect; the connection is made.

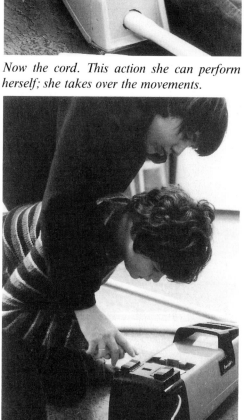

◀

Back to the vacuum cleaner. The switch is turned on; the resistance is changed – cause! And the effect?

A loud noise! And then?
It sucks at the face;

it sucks at the hair and ears.

And now look! What happens?
A piece of paper is sucked in.

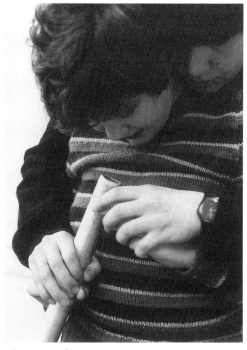

The paper disappears.

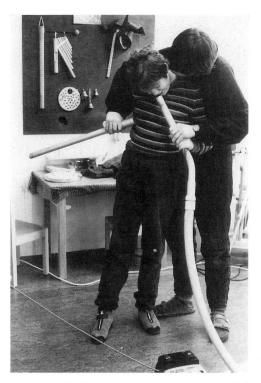

"It was right there! Now it's gone! Where did it go"?

"It is not in the tube! Is it in the vacuum cleaner"?

◀

"Watch"! The resistance is yielding and it can be opened. What one sees, one can touch, embrace, and take off – take out.
A whole chain of changes in resistance – of causes and effects – is present!

Here is the bag containing the dirt. "What is inside"?

"Can I touch it"?

"Embrace it? Take it off? Take it out? Move it?
And look, here is the piece of paper which had just disappeared.
What a funny thing"! A. laughs.

Now she can reinsert the sweeper bag. Again, a whole chain of successive changes in resistance – of causes and effects – occurs!

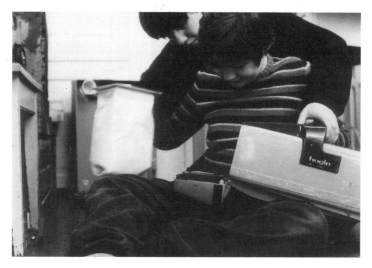

▲
The bag is in and the tube can be connected. There are changes in resistance again and again!
▼

▲
The noise is back; the pieces of paper on the floor disappear. The goal is attained; the floor is cleaned.

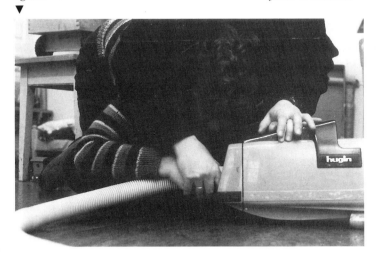

2.3.1 When to Use Hands and When to Use Tools

When I guide, as often as possible, I try to use *the body* for causing the changes that are needed to solve the problem. Most of the time, I use the hands; perhaps I can also use the mouth, the legs, or the trunk. Only when I do not get the desired effect by using parts of the body, do I consider the use of tools. This basic principle is important when I guide children or adults who lack information about their surroundings because of their perceptual problems. Here are a few examples of how we can use the *hands* in a direct manner.

A., 9 years and perceptually disordered, is squeezing a lemon. She is being guided.

The right hand holds the slippery lemon and presses it on the stable support.
The left hand touches the pulpy, moist part of the lemon.

Her fingers have made a hole. The opening becomes wider. She presses again into the soft, fleshy part; then she attempts to take it out!

Her fingers press into it. "Is the resistance yielding? Can what I touch be taken off"? Her fingers get wet. It smells! The skin is tough, but the fleshy part yields.

Success! A piece comes off and is lying on the plate. ▶

The *success of movements for causing a desired effect depends*, among other factors, on the knowledge about the *kind of material* I am working with. Therefore, in order to act adequately, one needs information about the material. In the following example, E. has to explore the quality of the pulp and seeds of a melon before he can take them out.

E., 3 years, 6 months and a normal child, is busy preparing a melon. The melon is peeled. Now he must take the seeds out. His mother guides his hands as soon as he cannot perform the actions by himself.

He touches the pulp of the melon. It is so soft and slippery! Mother and E. are smiling. Yes, these are the qualities of a melon!

Now inside!
He feels the resistance at the side, and he feels the soft pulpy part inside. It is moist and sticky! He takes over the movement. He is tense – even his tongue is part of the action!

211

The goal is attained! ▲
The seeds are on the plate. They have been taken out; they are separated from the melon. The desired change is performed.

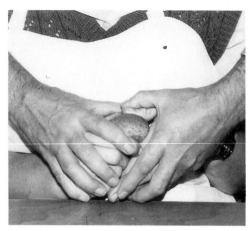

Her fingers embrace the roll...

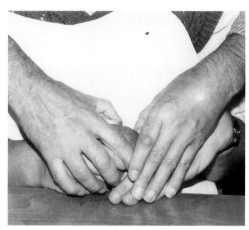

and search for the best position...

Breaking bread or a roll also requires exploration of a substance. *How* do I embrace the roll? *How* do I hold it in order to execute the required pressure? And *how* strong does the pressure have to be in order to break the roll?

F., 23 years and both perceptually disordered and cerebral palsied, is supposed to break a roll in half. (See also previous examples of cutting a cucumber, pp. 199–201)

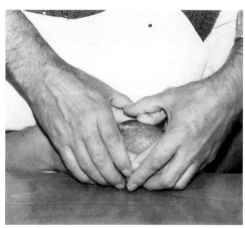

for breaking the resistance.

Now, in this position she can press firmly against the resistance of the support; the resistance of the crust yields. The roll is divided.

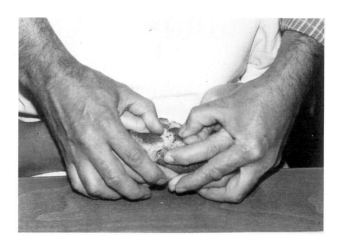

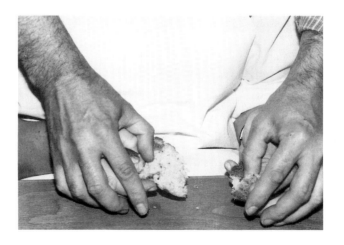

In the next example, F. is breaking a piece of cheese to put on a slice of bread.

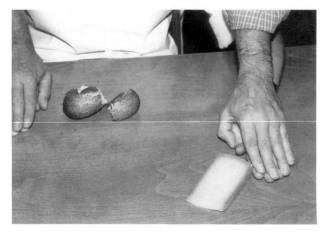

F. reaches for the cheese across the support

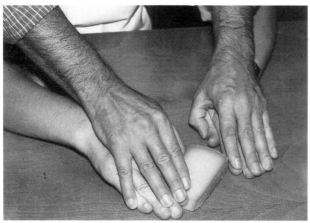

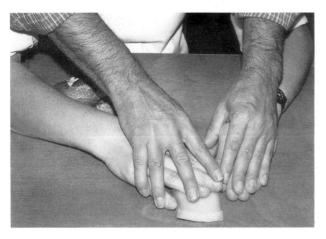

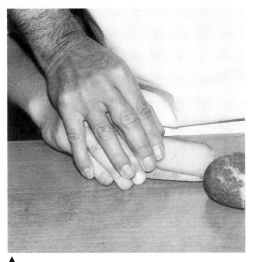

▲
and brings it closer to her body until she feels its resistance.

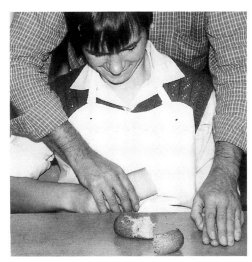

▶

With both hands, she presses the cheese against the resistance of her body.

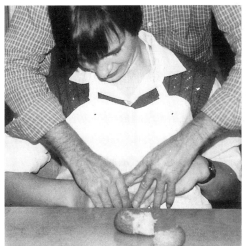

She attempts to break the cheese. ▼
She is successful. A small piece of cheese is embraced by her left hand.

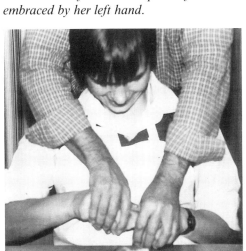

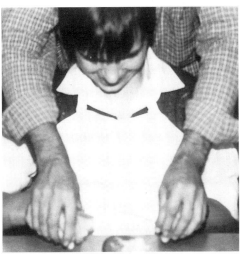

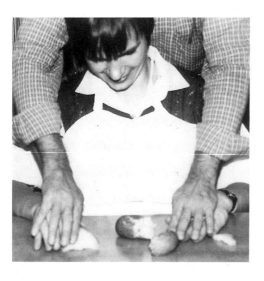

With her right hand, she puts the cheese she does not need back on the table and releases it there.

◀

Often we do not succeed by using only the hands. In these cases, we have to search for a *tool*, e. g., a knife, a spoon, or a hammer.

I am guiding C., a normal adult, in a practice session. We are holding an eggplant in both hands. The goal of our action is to prepare eggplant for supper. The objective is to cut up the vegetable!

With both hands, we press the eggplant on the support. Can we divide it in this manner – by using only our hands?

"No". We are not successful. We search for a tool – the knife. "Where is it? There. How do I get it over here"?

We touch the knife and slide it over the edge of the table to embrace it.

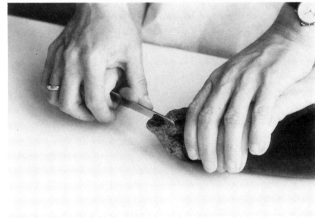

With the knife in the right hand, the eggplant can be touched.

The left hand touches the knife, too, while pressing the eggplant on the support. Now the cutting can begin.

D., 5 years and a normal child, continues to peel a melon (see the example about D. in Part III B, paragraph 1.1 "I Learn from Tactile-Kinesthetic Information"). He uses a knife from the kitchen.

▲
His mother takes his hands and continues the way she would do it herself – in harmony, without interruption.
D. doesn't notice he is being guided. He focuses his attention entirely on the task, the peeling.

He can feel the slight pressure of the knife cutting along the resistance of the melon.

◀
He feels the pressure of the melon he is holding in one hand becoming stronger. Now, the resistance against his hand holding the knife suddenly yields, and soon the resistance becomes total. The knife has touched the plate and a piece of the rind is cut off.

In this manner, we directly cause movements with our bodies or with the help of a tool. We try to reach a desired change. When solving problems in daily life, we should examine such effects over and over again. Are they as we want them? The examination should always include changes in resistance.

A., 9 years and perceptually disordered, cores an apple in order to put a filling in it. She is being guided.

To remove the core, she takes a knife and pushes it into the apple. Doing so, she holds the apple against her body and presses it upon the table. In this manner she can feel the changes in resistance when pushing the knife into the apple. The knife is deep inside the apple and must be turned! Pieces of the apple come out and make a hole.

She pokes inside the opening with her finger.
What is the effect? Is the hole large enough? Her fingers explore it and find that it is still too small.
She continues to poke and scratch more of the pulp away with her fingers.

She puts her finger through the hole, feeling it. Then she looks through it. The hole is very big now!

2.3.2 Everything Happens in Its Own Time

We must be aware that causes and effects follow each other in a time sequence. Sometimes the sequence can be determined *optionally*. When I prepare a melon, I can first peel it and afterward, divide it – or conversely – I can first divide the melon and then peel it. The cucumber and eggplant, can usually be cut into pieces first, and then the pieces can be put into a pan. However, it might also be possible to put the cucumber and eggplant into the pan first, and cut them up afterward.

Besides the sequences of actions chosen from options, there are sequences which are *constrained*. In order to take seeds out of a melon, I have to cut open the melon first. When filling an apple, I have to make a hole in it first. Obviously, there are constraints on the sequences in the example of the vacuum cleaner. First, I must assemble the parts and then turn it on.

In this sense, problem solving events provide a frame for the meaningfulness to sequences of causes and effects, and thus, are a fundamental part of learning and development.

2.3.3 Am I Allowed to Break That?

Many actions we perform cannot be reversed. This can result in serious consequences. Think, for instance, about the effect of polluting the environment!

Children like to explore their surroundings. They want to know what things are made of. Such explorations often lead to *destruction*: A watch is taken apart or a doll's eyes are pushed in. In *daily events*, destruction is possible and can even be necessary to solve a problem. Such events include *changes, which are not reversible*. The following example illustrates this:

B., 5 years old and perceptually disordered, comes for an evaluation. He is described as hyperactive and unable to sit quietly, even for half a minute. Nothing is safe around him. Whatever he touches, breaks.

He is guided in preparing an "apple in a crust". Toward the end of the preparation, an egg has to be broken so that the egg yolk can be used to coat the crust. Several times he repeats, "Egg broken! Egg broken"! Even after the task is completed, he comes back to the incident of the broken egg.

Almost all events in the kitchen are characterized by having to break something. The kitchen, therefore, is an ideal place for children to experience events having irreversible changes.

To Summarize:

In this chapter, "I Feel and Act Upon," we have discussed the *strong relationship between the qualities of the surroundings and our acting upon them.*

In order to act upon something, I must feel the quality of the material. This is made possible by touching and embracing something that moves in reference to a stable support.

To receive enough tactile-kinesthetic information, I directly use my whole body, if possible. I will use a tool only when I am not successful in using parts of my body.

I can *only get tactile-kinesthetic input when my body is moving or is moved*. I can feel resistance only when I change resistances. Changes always have to occur relative to a stable reference surface.

The *stable support is vital for picking up such information*. We repeat: The stable support serves as a reference for the changes in resistance. Thus it is a base for adequate touching, embracing, and releasing, as well as for exploring neighboring relationships.

When I am successful in transmitting information to the perceptually disordered about changes in resistance by guiding them through problem solving events, they will experience sequences of causes and effects. They will also receive tactile-kinesthetic information about materials, and changes of the Wirklichkeit within these events.

3 I Understand Problem Solving Events of Daily Living

In the first two chapters of Part III B we described how we can elicit changes in behavior by providing tactile-kinesthetic information.

We described the differences among the four levels of learning: jerking away, becoming familiar, recognizing, and anticipating. However, even with these four levels of learning we have not yet arrived at a level of production. Performances of production will be discussed in Part III C. In this third chapter we will continue to discuss the levels of learning with emphasis on levels 2 through 4. *Learning* occurs only when the learner's "state of mind" is present. In this regard, the *concept of understanding* will be central.

3.1 Learning Begins With Understanding

Understanding is a prerequisite for learning. If I do not understand something, I cannot store it. Therefore, I cannot learn. What do I mean by understanding? How do I recognize when understanding is present?

3.1.1 Understanding – What Is Meant?

When I guide someone through a problem solving event, I can feel if the person being guided is "with the event," that is, involved in it. I can feel if the one being guided shows the same changes of tenseness that I do. Each of us relaxes briefly on the support – our tension decreases. Each of us moves. My tension increases and so does the tension of the person I am guiding. When I move in the direction of a needed object, the person I am guiding follows. A strong tactile-kinesthetic relationship is established between him or her and myself, and between us and what is happening in our surroundings as caused by us. We often use a Latin expression to demonstrate this point. The Latin *inter-esse*, i.e., being in between or being part of, occurs if a person becomes a part of the event. When this occurs, there is *understanding*. It is an important concept. We can also call it the "state of mind" a person has in regard to an event.

We appear to be able to use a kind of "antenna" in judging whether a person shows understanding or not. There are some – and I fear they are becoming fewer and fewer in number – who, just by their presence give others the feeling of being understood. I had an aunt who lived in one of the mountain valleys. She spent most of her time in her big family kitchen cooking for her large family. She was the heart of her family, but the heart of the whole neighborhood as well. While she worked in her kitchen, anyone could come and sit down at the table. Even if she appeared to be busy with her work, one always had the impression that she "is present to me," meaning she was involved with me. Even if nothing was said, one had the feeling that she "understood".

I often teach postgraduate courses. The participants have taken time off from work in order to learn. To fulfill their needs, I have to remember that learning begins with understanding. How do I know if I am fulfilling this condition for them? By what characteristics do I recognize that the participants understand my presentations? At the beginning of a course, I know very few of the participants. Therefore, I do not know their level of knowledge. I cannot infer it from their production level because I know that understanding comes before production and is much more encompassing.

I cannot evaluate understanding in a direct manner, as can be done with production. I can only observe the behavior of the students. When I observe specific kinds of behavioral patterns, I infer: Now the students are following my presentation. Now they understand what I am saying or what I am doing. What are these behavioral patterns? At first, there is a certain tenseness expressed by the behavior of the participants. By this, I mean a positive kind of tenseness – in contrast to a negative tenseness mentioned earlier, (see Part II, paragraph 2.1.1 "They Withdraw From Touching, Become Tense and Look Away"). I am talking about an alertness which I can observe, for instance, as I tell children a story. They listen attentively and with understanding. Likewise, effective speakers can elicit this kind of readiness in their listeners. It is expressed by a certain increase in body tone. When people are tense, they do not sit in a relaxed manner in a chair. Tenseness is also observable. I can see it in their faces – the eyes are slightly enlarged and unnecessary movements are suppressed. As a speaker, whenever I observe this behavior, I infer that I have elicited understanding in my listeners.

To evaluate if my listeners show understanding I can also observe the behavioral patterns previously described in the discussion about the levels of learning, 2 through 4: I observe if they are looking at where the events occur, in our case at me, the speaker – just like children look at the actions performed by their hands. I watch out for signs of recognition. And are there continuing movements? Do they show anticipation?

3.1.2 I Work with Children and Adults on Their Level of Understanding

When I teach a course, I organize the content according to the level of the participants' understanding and not according to their level of production. They apply for the course because they feel they cannot meet the expectations of their patients. They want to know more, i.e., to extend their understanding. This is their immediate goal. Their ultimate goal is to have better production for their patients.

In a similar way, I plan my work with perceptually disordered children and adults. To enhance progress as intensively as possible, I have to orient my work according to their level of understanding, not their level of production. I intend to provide them with as much adequate tactile-kinesthetic information as possible – information they are not able to receive when left alone. For instance, for working on problems they can solve in a familiar situation I choose a less familiar one.

To illustrate this point the following examples are taken from a field trip:

C., 14 years and perceptually disordered, is supposed to cut zucchini, but how different the situation is as he sits on a rock and not on a chair in the kitchen. C. is being guided.

He balances on the rock, searching for the stable resistance of the support.

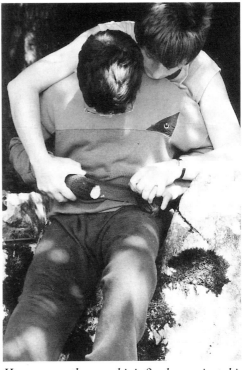

He presses the zucchini firmly against his body and then begins with the knife. He succeeds in cutting one piece.

◀

On a plate he cuts it into smaller pieces. Again he has to search for a stable resistance. There is the chance that the plate will slide off the left side.

Now the support is stable. The task can be completed.

Another problem: The sausage for preparing a hot dog needs to be put on a stick. How should he do that?

◀ *P., 15 years and perceptually disordered, helps to prepare the campfire for cooking. He is being guided.*

Breaking branches for the fire is not an easy task. There are many changes in resistance he has to perceive!

"How do I hold the different branches? How much pressure do I have to apply"?

For cooking, the pan has to hang over the fire. "How do I put water for the soup into the pan"?

The soup is finished cooking, and the two pans can be taken off the fire and stick.

The first pan is at the end of the stick. "Be careful! The pan is hot"!

He sets the pan on the grass next to the fire. "Be careful! Don't let it spill"! Now he must remove the second pan.

◀

Here – the second pan is taken off the stick. "What comes next"? The pan is in one hand, and the stick is in the other one.

In the discussion of the failures (see Part II, paragraph 3.3.4 "…and the Competence Does Not Become Performance"), we differentiated between competence and performance. I can or cannot perform. In a familiar situation, I can perform many actions. When the situation becomes unfamiliar, performance breaks down. This is the case for us as normal people. It also happens to those who are perceptually disordered, but because they lack information about the changes of situations, it happens sooner and, consequently, more often.

With regard to the above examples of the field trip, C. had cut zucchini many times before, and it was not new for P. to break things, such as twigs, or to hang things up. For both boys, though, the actual *situations* were strange and *unfamiliar*: to sit on a rock and balance on it; to stoop down and break twigs in a new position; and to hang a pan on a long stick. Also, *the materials* were unfamiliar. The zucchini had to be cut on a paper plate, supported only by the knees; a tangled mix-up of branches had to be broken; the sticks to hold the two pans over the fire had to be steadied on two round stones. As a result, the two boys would not have been able to perform had they been on their own. To solve the problems, each needed tactile-kinesthetic information about the characteristics of the new situation, i. e., information about the materials being used and the necessary causes and effects. Guiding provided more adequate tactile-kinesthetic information. We observed how the eyes of both boys followed what was happening with their hands and bodies. They both showed *understanding*, a prerequisite for learning! This is the level one should work for with children and adults.

It is not always easy to find the level of understanding, especially with children or adults I do not know. When I observe them performing spontaneously, I get information about their level of production but not about the extent of their understanding. I only know one thing: *Their understanding is always much more comprehensive than their production*. This is also valid for perceptually disordered children and adults. I must always consider this fact when I prepare a problem solving event for working with a perceptually disordered child or adult.

M., 5 years, is perceptually disordered. Therapists report that he can concentrate on an activity for no longer than half a minute. He touches objects, but hardly ever grasps them. He seems to be restless.

M. is being guided in the preparation of chocolate pudding. We open the milk carton, explore the opening, and pour the milk. We open the bag with the chocolate pudding mix – touching, embracing, exploring it. We pour it into the milk, grasp the whisk – touching, embracing, and moving it; we stir the mixture, and taste the chocolate pudding.

For 20 minutes, he shows full understanding of the event. Now, I lift him off my lap, putting him down. He climbs back on my lap. It is obvious he wants to continue the task.

The title of this chapter is "I Understand Problem Solving Events of Daily Living". After having considered what understanding means, we will briefly discuss the concept of "problem".

3.2 Problems Arise All the Time

Daily living situations are always changing. The changes in the situations continually create problems. Besides these unintended problems, there are problems which we create voluntarily. Children are especially skilled in discovering these kinds of problems. They search for them – openly and with enthusiasm. We will briefly discuss this amazing phenomenon.

3.2.1 Problem Solving Is Thrilling

Events of daily living involve problems. When I observe the behavior of children in their daily lives, I become conscious of their continual search for problems which are part of an actual situation, e.g., a wall to climb on, a fence to slip through, a hole to hide in! How boring it is to perform an activity which does not include some difficulties! Children want to solve problems, to master difficulties, and to take on challenges. Isn't this also the case with us as adults? Perceptually disordered children and adults are no different from us in this regard.

To help impress upon you the importance of this behavior, we will add some more examples observed in daily living situations.

M., 8 years and a normal child, climbs a tree stump.

What an effort! He is sweating. Climbing is very difficult. He finally reaches the top, sits on the stump, and looks down at us – proud that he has solved that problem.

J., 4 years, 10 months and a normal child, watches M. climb up on the tree stump. She is standing next to a rock which is one-third her size. She discovers the rock and climbs on it. Her feet no longer touch the ground. She looks around and comments, "This is dangerous".

The first snow of the winter season has arrived. The road is partly covered with ice. A boy, approximately 10 years old, is sliding on the ice, able to keep his balance, but with some effort. As he passes me, he looks at me and proudly declares, "Today it is dangerous"!

M., 18 months and a normal child, discovers three glasses on the table.

She has just finished piling the three glasses on top of each other. She touches them and presses on them with one hand. "Is the resistance a stable one"?

Cautiously she takes the glass tower apart. Using two hands, she embraces and puts one of the glasses upside down on the table. She ponders the situation.

Once again, she constructs a tower. She turns the third glass over and puts it on top of the second one. She explores the resistance. Her mouth is open because this task requires much attention!

▶

Another trial is successful. The glasses are inside and above each other. "Does the tower hold"? M. looks at her construction with complete concentration.
◀

Could she construct the tower another time? Intensely, she begins her work again. "Solving problems is so thrilling"!

And now, she tries a different construction! There are always changes in resistance! She is constantly feeling and looking!

"Look – a new way"! The glasses rest on their tops and on top of each other! "Will the tower hold"?

Could she risk lifting this tower off its support?

Could she place the glasses on each other? Into each other? Which is greater? Which is smaller? Qualities of material, neighboring relationships, and their interdependence are all explored through changes in resistance, thanks to the tactile-kinesthetic system.

M., 2 years, 5 months, is on a hike with her father.

There is a creek with many stones and a pool of water!

There are also some stones on the road. She can take them off and throw them into the water.

"How many can I hold in my hand"?

Closer to the water, it is even more interesting. There are many, many more stones to be picked up.

Now she climbs up on the large, flat rock. Climbing is difficult with so many stones in her hand.

"Now, watch"! The stones fly through the air into the water. "What a splash! What an effect"! ▼

3.2.2 Difficulties Can Be Overcome; How Good It Is to Have Difficulties

Depending on the situation and the complexity of the tasks, solving problems is more or less difficult. Difficulties which naturally arise, can help to increase the thrill of problem solving. Children demonstrate an astonishing skill in choosing a problem that corresponds with their abilities. M. in our example chose a tree stump which was quite high and J. chose a low rock. Each one chose according to his or her ability to solve the problem. Each was successful – and proud – at having mastered a difficulty.

Those who are perceptually disordered hardly ever experience this kind of success. As we observe them in their spontaneous activity, we note how they panic, become agitated, or begin to talk endlessly when difficulties arise. In Part II, paragraph 1.2 "They Talk Incessantly," we referred to this kind of behavior. We interpreted the exhibited behavior as being a sign of their failure caused by their lack of tactile-kinesthetic information. If such a state lasts, they could become obstinate or withdraw and avoid situations with people moving around.

It is important that we acknowledge difficulties which appear during daily problem solving events when we are guiding someone. Unskilled actions should not be avoided. We should not be afraid when an unforeseen event occurs. Guided problem solving events should represent daily living – life with its innumerable difficulties. I cut an apple. The knife slips out of my hand. A piece of the apple falls on the floor. The sleeve of my sweater, which I had just rolled up, slides down again. I forgot to put the bowl with the pieces of apple on the table. All of these are wonderful opportunities to solve the unexpected problem and master difficulties while guiding a person. How often we must touch and move. How many changes in resistance we can feel until the respective difficulty is overcome. And what an experience it is to reach that goal!

The following example concludes this paragraph:

This event happened during a course for a group of families. The fathers had to prepare a problem solving event to perform with their children. The father of R., 5 years old and perceptually disordered, was very nervous. He pointed out that he could not work with his hands and would undoubtedly be very clumsy. To prepare something to do by hand – even cutting – would be very difficult. Finally, though, he agreed to try.

The goal was to prepare a small surprise bag for R.'s mother. In the middle of the task, the staples ran out. This was not part of the plan.

With courage, R.'s father kept on guiding him, and those of us who were watching held our breath.

The father continued – even with all his insecurity and difficulty in finding out how to open the stapler, fill it, and close it afterward. It was also difficult to get enough information to perform all of these steps of touching, embracing, and moving. After having turned the instrument over to touch, embrace, and move it – again and again – the stapler was finally ready to function and the planned activities could continue.

In the end, not only did the father take a deep breath, but R. did, too. It seemed as if neither of them had had time to breathe during the problem solving activity.

This guided event became the high point of the day. The other fathers, even those who were more skilled, were unable to elicit as much attention from their children as this unskilled father. Because of mastering difficulties together and because of mutual touching and feeling, this was quite an experience!

When we observe normal children in their multitude of problem solving activities, we are surprised by another aspect.

3.2.3 The Solving of the Problem Is Important, Not the Product

M., 19 months and a normal child, is busy gluing pieces of paper on a cardboard. As soon as she sees me she laughs and comments, "Ike luing" (like gluing). The products on the pieces of paper are being put on sideways, but this does not matter to her, even after she is all finished.

We can make similar observations with perceptually disordered children.

A., 8 years and perceptually disordered, prepares an apple tart with her therapist. She is attentive but at the completion of the preparations, she doesn't even ask for it.

However, perceptually disordered children and adults can often exhibit a behavior that does not correspond to that described in the above examples. The product becomes more important than the solving activities. When this occurs they usually judge their product as being "ugly" or "not right". They seem to compare to a kind of preconceived picture they have of the object – a picture which, similar to a photograph, reproduces visual information about the object.

A., 8 years and perceptually disordered, is being guided in preparing a ham croissant.
The goal is reached: The ham is on a piece of dough, the dough is rolled around the ham, and the whole thing has been baked. A. takes it home. At home, though, he complains that it is not like the ham croissant you purchase at the bakery.

If we consider the visual "appearance" of A.'s ham croissant, his judgment was correct. His ham croissant was not as nicely rounded as the ones from the bakery, but his activity of making a ham croissant was correct. The ham was rolled into the dough, just like the one from the bakery. It is strange that A. has focused on the visual impression to judge his product. To concentrate too much on what can be seen in these situations, means that a child or an adult is oriented too much toward visual information, and lacks tactile-kinesthetic information. This is different from the behavior of a normal child as shown in the following example:

T., 5 years, is on vacation with me. I am sewing. T. wants to sew, too. I give him some material, and he cuts and sews, in any direction. By the time he is finished, several pieces of material are attached to each other, forming a long ribbon. He contemplates his work and comments, "a belt". Later on, as he travels home on the train, he wears the fringy, unskillfully cut and sewn, long piece of material around his tummy. Beautiful to look at? No! But how does it feel? Exactly like a belt.

3.2.4 I Have Solved the Problem by Myself

When people watch the performance of a guided event, they often ask, "Don't these children and adults become passive when you guide them? Don't they get the impression that the person who guides them is solving the problem and not they themselves? Don't they start to let those around them in their environment solve their problems"?

As was mentioned before, normal children can also be guided under the condition that the task is too difficult for them to perform by themselves. While they are being guided, their change of facial expressions indicate whether or not they have received tactile-kinesthetic information in this situation similar to the kind they would receive if they were to produce the task themselves. This means that they perceive the elicited changes in resistance under both conditions – being guided or producing it themselves. It appears that each time tactile-kinesthetic information is received during the solving of a problem, they reach the conclusion, "I have done it by myself".

We can observe the same phenomenon in those who are perceptually disordered.

Ch., 5 years and perceptually disordered, is making a pizza. He is being guided through the task. No talking occurs during the guiding.
After finishing the pizza, he runs to his mother and tells her, full of joy, "I made a pizza"!

Among the variety of conditions needed to create the impression, "I have done this myself"! when being guided, *two conditions* are of special importance. We must guide through all the steps, which the child or adult cannot perform spontaneously, and we should not talk as long as they pay attention.

The following example illustrates these conditions:

R., 10 years, is perceptually disordered. His father is making a wood carrier with him so he can carry wood from the cellar to the fireplace upstairs. While guiding, the father names every step but performs some of the steps by himself, without guiding R. In these situations, R. watches his father. At other times, R. performs some actions which are habits for him. At the end we ask R. what he has done. He answers, "I helped my father".

This situation occurs during a course with families. Afterward, we discuss the method of guiding used by the father and suggest changes.

The father applies the changes to another attempt at making a wood carrier. He chooses problem solving activities which have no habit characteristics for R. For instance, during the work, he makes the manipulations more difficult by having R. take on different hand and body positions and by having him use other kinds of tools. He guides R. through all the steps and does not talk to him. At the end of the session, we again ask R. what he did. He proudly answers, "I made a wood carrier"!

It appears that when I guide and want the guided person to receive the impression, "I am the one who has solved the problem," it is necessary that the guiding includes all the required steps (except when he or she can produce the activity spontaneously) and that I not talk as long as there is attention. In considering these conditions you could ask, "If I cannot talk during the problem solving event, when can I talk? When should I talk"?

3.2.5 When Do I Talk?

Solving problems requires tactile-kinesthetic information. Therefore, it is important to pay full attention to the tactile-kinesthetic input during the performance of an event. Talking becomes unimportant; it can even be obstructive. It distracts attention from the tactile-kinesthetic input and disrupts the guiding. Let us put ourselves into a critical situation where our attention is focused entirely on the tactile-kinesthetic input. I may need to pour something into a container with a very small opening and I do not want to spill any. If someone is persistently talking to me at my side, I will either ask that person to be quiet and wait until I have finished, or I will put my task aside and switch my attention to him or her.

The same is valid when guiding children or adults through a problem solving task. I briefly *speak* with them *before* beginning the activities. For instance, I tell them what we will be doing. My choice of words depends on their level of verbal comprehension, but once the activities have begun, I do not talk for as long as their attention is directed on the event. Furthermore, I am not able to talk since I, too, have to focus my attention on the guided event.

If a mishap of some kind occurs during the guiding and, as a result, *tension increases* and understanding breaks down, I interrupt my guiding and try to decrease the tension level by talking. This might also help decrease my tension, which is usually just as high.

I will *talk after* I have finished guiding through an event (see Part III C, paragraph 2.4.2 "First Comes Tactile-Kinesthetic Input from Problem Solving Events of Daily Living and Then Comes Representation"). This is by far the best time for representing the event which has been perceived through the tactile-kinesthetic system. To do this, I can use drawings and/or verbal or written forms. What I choose, depends on the level of understanding of the ones being guided. Since it concerns *understanding*, I choose a degree of difficulty and form that the guided persons are incapable of producing by themselves. Under these conditions, *I speak, draw, or write*, and I expect that the perceptually disordered children or adults will listen and watch, just like normal, young children do who are not yet able to draw or write.

We spoke about the importance of understanding in order to learn and how thrilling it can be to solve problems. In this discussion "daily living" was mentioned quite often. In the next paragraph we will consider that concept more thoroughly.

3.3 Guided Through Daily Living

The problems normal children discover and try to solve are related to situations in daily living. They are thus part of the Wirklichkeit: The tree stump that M. climbed on was located next to a rest area; the drinking glasses that M. used for constructing towers were on the table; the ice the boy slid on covered the road on the way to school.

3.3.1 I Guide When Problems Arise in Daily Living . . .

Daily living situations offer opportunities to solve innumerable problems. This is also true for those who are perceptually disordered. Their difficulties are most obviously expressed by failures in daily living events (see Part II, "Failing in a Wirklichkeit"). Daily living situations offer excellent opportunities for guiding perceptually disordered children and adults through problem solving events, and they thus, provide them with meaningful tactile-kinesthetic information. An example might be something as simple as opening a door. When a handicapped person wants to go through a door, some hurry to help - open the door, let the person go through, and close the door afterward. Wasn't that being really nice and helpful? No! An excellent opportunity to have a guided situation was missed, and an ideal chance to help the handicapped person advance was overlooked. It would not have taken any longer to have done the following:

- Take the hand of the perceptually disordered child or adult and perform with him or her all of the necessary movements: Open the door, walk through it, and close it afterward.

Imagine the procedure if he or she is carrying a cup of milk. Guiding the person, we put the cup on the floor in order to have the hands free. We touch the door with one hand - feeling its resistance. We embrace the doorknob and turn it with the other hand, pull the door towards our bodies and squeeze through the opening - paying attention to the different changes in resistance, always searching for a stable one. In this event, the guided person experiences causes and effects in a sequence - touching, embracing, and acting upon - and the solving of problems. When do I close the door again? When can I take the cup of milk off the floor? When this step is reached, the child or adult can follow what he or she intended to do, and I end my guiding.

In this way I provide children or adults who are perceptually disordered with precious tactile-kinesthetic information within a meaningful task for solving a problem in daily living.

Often there are opportunities in which one of the parents, therapists, or educators is already standing nearby, e. g., getting dressed, preparing to eat, pouring something into a cup, preparing a simple sandwich, adding cocoa to milk, opening a yogurt, or needing a napkin. In the case of needing a napkin, for instance, the parent, therapist, or educator would have to stand up, walk to the closet, open it, find the box with the napkins, and take one out of the box anyway. What an excellent opportunity to take the person's hands and guide him or her through tasks like these. It will only take a few moments of devoted attention with the child or adult, but it will give him or her an important instance of tactile kinesthetic information - information connected with the experience of having solved a problem of daily living - instead of having failed again.

To become conscious of such opportunities - and this can be quite difficult - we have to change our behavior so that *not caring for, but "going with,"* is our motive. I am not here for the child or adult, but with the person. We will only make progress in this way.

K., 10 years and perceptually disordered, is standing in front of an open closet. She wants to get the hot plate, but it is way up on the top shelf.

She looks up, thinking over the problem. ▶

▼

A teacher who is near by recognizes the situation. She begins to guide K. to solve the problem.

Up - she guides the arms to reach up high - searching for resistance, finding it, and grasping a shelf. Now K. should move her feet.

▲
K. does not know where to put her feet. The teacher guides her legs, exploring the resistances.

Finally she is high enough to touch and embrace the hot plate; she can take it off the shelf.

▶

She must come down with the hot plate in her hand – this is difficult!
The teacher guides K. in exploring the changes in resistance until she reaches the next shelf.
Now her whole body is moving. Where is the next stable resistance?

Each foot is on a different shelf – each one feels a stable resistance. In this position the hot plate can be released, and for a brief moment, K. feels the stability of her surroundings with different parts of her body.
◄

▼
K. is on the floor, both feet on a stable support! Now she can move the hot plate further down.

How can she best embrace the hot plate? Where are the changes in resistance?

▶

The problem is solved. K. has had an important experience: She has touched, embraced, and moved; she has received tactile-kinesthetic information about effects; she has understood their causes and reached the goal.

Taking a spontaneous opportunity to guide in daily living when we are near by, may be called "incidental guiding". The *tactile-kinesthetic information* we provide in these moments is *very precious* because it is *related to the solving of a problem which is connected in a meaningful way to the Wirklichkeit of daily living.*

Incidental guiding requires little time – and most often, the time would have been spent doing a task for the person anyway.

Unfortunately, therapists have to work with a child or adult at a specific time, and then, only for a limited amount of time. Therefore, they must pack as much as possible into their therapy time; they cannot wait until a problem in daily living arises. They must prepare problem solving events – plan them.

3.3.2 ...or I Have to Plan Problem Solving Events

Several circumstances make it difficult for therapists to include the Wirklichkeit in their therapeutic work.

There is the room situation: Therapy rooms are often in school buildings or in clinics where construction, arrangement and furniture have little resemblance to the patients' homes. The highly centralized daily routines of such places aggravates this situation. Meals come from a central kitchen. The preparation of meals and whatever is related to them, e.g., planning, shopping, preparing and cooking are taken from the field of action of the children and adults involved. The wash is collected in a central place where it is cleaned, repaired, ironed, folded, and distributed. Living areas and school/therapy/working areas are close together. The problem of, "How do I get from one place to another"? hardly ever arises.

Gardeners and other specialized laborers, are hired to take care of the landscape and buildings. Lawns are mowed by big machines.

For therapists in institutions such as this, finding meaningful problems in the living situations of the patient becomes a difficult task. What can one do under such highly organized circumstances?

One possibility for including the patients Wirklichkeit despite the difficulties, is to try to be in communication with their different "worlds": therapy, classroom, ward, educational group, and *especially, and above all others, the family*! The stronger the interrelationships among those different worlds, the easier it will be to take actual problems into consideration. In other words, the better able I am to plan problem solving activities in daily living situations, the better it will be for the patient.

- There is a birthday in the family. The child or adult could prepare a small pastry to be baked in the oven at home.
- An object from home is broken. We repair it during therapy.
- Somebody in the home would like to plant a flower in a pot. This can be done in therapy.
- A shelf is dirty in the classroom or on the ward. We clean it with the patient.
- A glass with nuts is empty at home. We fill it during therapy.
- The mother wants to cook a vegetable soup at home. Onions and carrots should be cleaned and cut. We can do this in therapy.

Other possibilities include *changing the environment* as much as is feasible: Arrange a cooking niche so that small meals can be prepared; decrease the centrally organized activities so that parts of those events can be included in therapy; prepare coffee for the group of therapists; wash the dirty dishes; clean out the sink in the patient's room on the ward; and change therapy time so that it includes meal preparation and eating, getting dressed, or getting ready for bed.

3.3.3 Feeling Is Very Difficult

Even when we have situations where problems arise for guiding children and adults in their daily living, it is often difficult to accomplish the guiding.

The main reason for this difficulty is the hectic life we lead as part of our own social groups. We are involved in our own daily situations which create many problems. We have to try to solve them. Therefore, our capacity to respond adequately in a guided problem solving situation is often highly reduced.

It is under these conditions, though, that I should take time to consider the problems in daily living that a child or adult has – problems which are not of my own, not of a personal kind.

Even when I have discovered what the problems of a child or adult are, I still must be ready to guide. I have to switch to a tactile-kinesthetic input! This is very difficult in our culture which is tuned toward seeing and hearing but not feeling. Our eyes are blinded by the intensive lights of advertisements everywhere. We become intoxicated by noise all around us. We can no longer stand silence. Wherever we go, there is music in the background: in restaurants, in department stores, at swimming pools, and at home. Feeling has been pushed aside because we are so involved in seeing and hearing.

Switching to tactile-kinesthetic input demands some special attention and effort each time. Often, the time needed to make the switch is too long. The child or adult has already received the wrong kind of information – may even have become conscious of failing again. The panic is there! They may already be attempting to cover up their failure through strange behavior. When this occurs, it is difficult to guide them.

Guiding is filled with feeling, and therefore, I usually tire more easily and more quickly than the person who is being guided. I need breaks. This aspect is not always considered.

3.3.4 A Break in Guiding Allows Time for Thinking

The first kind of break is called a *breath break*. It is short – just long enough to allow me to take a breath. It should be used frequently. It is similar to pausing after speaking a sentence. I have finished a step, e. g., we have grasped the knife we need, can finally embrace the banana, have reached the refrigerator, or taken off the shoe – now we can breathe again. While doing this, we may hold the grasped object against a stable support. A *stable reference surface is important for being able to breathe*. We cannot do this while the surroundings we touch are moving. It is different when the surroundings resist our movements. Now I can stand up, and the guided person can stand with me! I can breathe, relax, think over what has happened, and decide what comes next!

The breath break also allows me to *feel the guided person as intensively as possible*. Are my fingers on his or her fingers? Is he or she relaxing, too? Embracing the object like I do? Feeling its resistance like I do?

Or was I not successful in transmitting changes in resistance to the one being guided? Did I guide through the air? Because of that, did the person become hectic? In this case, the breath break will not elicit relaxation, but on the contrary, it will increase tension. If it does, I have to decide whether or not to interrupt the guiding and insert a longer break – the *interruption break*.

The interruption break should give me time to relax, to be quiet, and to rebuild forces and energies. Frequently, this is a time I can ask the child or adult to perform an activity which belongs to the event in a meaningful way and which they can do by themselves, e. g., to carry the dirty dishes into the kitchen; to get a towel from the kitchen to clean the table; to wash the hands; or to put objects which are not used anymore back into the closet.

To Summarize:

In this Chapter 3 "I Understand Problem Solving Events of Daily Living," we further emphasized how important the level of understanding is for learning.

I do not work with students on their production level. Likewise, I don't work on a production level with perceptually disordered persons. I try to work with them on their level of understanding.

Understanding was defined as inter-est, as being in between. Understanding occurs when the children or adults are with the task, and I can transmit tactile-kinesthetic information to them within a guided problem solving event in daily living.

I choose problems from daily living situations. I keep in mind:

- Problem solving is thrilling.
- Difficulties exist in order to be mastered.
- I guide – without talking and with breaks – through each step.

When I work with those who are perceptually disordered in this way, what will happen to learning? What comes next? In Section C of Part III, we will try to answer this question.

C. Tactile-Kinesthetic Experiences with Solving Problems of Daily Living Are Interiorized

If you have worked a long time with children and adults who are perceptually disordered guiding them through problem solving events in daily living, you may remember it being quite difficult at first. You had to make a special effort to carefully touch things in the surroundings with them. You had to search for resistances with them. You had to embrace and search some more. You had to move and change resistances. Changes in resistance were again needed for releasing. You had to take your time in order to elicit all of these changes in resistance and in order to feel and appreciate them.

From the beginning, these children and adults began to show understanding as you guided them through problem solving events. Their hands and eyes began to coordinate while they were performing a task. Little by little their understanding widened. You observed them begin to show curiosity and seek information. They recognized the things they touched. They asked for more guided problem solving events. You could guide them for half an hour, even an hour. You became tired but they did not.

In the course of guiding, questions would often arise: Why all this work? What is your goal in guiding these people? Each time someone asks me questions like these, I respond, "It is like being a mountain hiker. Before us – the ones I am guiding and I – there is a huge mountain. For some of these children and adults the mountain is higher than it is for others. It doesn't matter, though, whether the mountain is higher or lower. The view from the top will be magnificent, and when we arrive there, we will have such a wonderful feeling of accomplishment. An example of this feeling follows:

I am on a hiking tour with M., 10 years old. We had gone on several long hikes during the last few days.
This is our last day and we are at a mountain creek. He begins to play and though I know he likes to do that, time is flying. We must decide if we should hike further or stay here and relax.
I ask him, "M., what shall we do? It's your last day. You choose".
He is silent.
After quietly thinking it over, he answers, "Let's hike further, over there". He points to a trail that goes up. We are on our way.
After some moments of silence, he asks, "Do you know why I wanted to hike further"? "No," I reply.
He explains, "When we reach the top, when we've done it, I will feel so good".

Yes, when you arrive at the top of the mountain, the feeling is good. We intend to reach this kind of feeling with the children and adults who are perceptually disordered. Having such a good feeling does not mean that we can live alone in a secluded place away from the world. Feeling good means being part of the Wirklichkeit, feeling what is around us, and meeting problems which we can solve. Feeling good means that we can master the difficulties along the way, recognize causes and effects, feel changes, and even elicit changes ourselves. There is an interaction of give and take. When this is the case, life is *meaningful.*

But, just as mountains are higher or lower, so also the goals of the individual children and adults are on different heights. When grown, some may perform meaningful work within a protected small group while others may be able to find their way independently and only need special help in troubled situations or when much is demanded of them. Others can fully integrate into the social reality. The height of the individual's mountain depends on several factors: the degree of the perceptual disturbance, the amount of tension within the surrounding group, the mental capacity of the individual, the possibility of schooling, the individual's health condition, and a variety of other factors.

The final goal is for the person to be able to lead a meaningful life. The way to reach that goal is a long one and requires many years of travel for all of us - the children, the adults, and ourselves. The next chapters will attempt to describe ongoing paths of travel.

1 Production Begins

1.1 From Anticipation to Producing

After a long period of increasing anticipations, the moment for producing actions arrives. Instead of looking at their fathers to cool the soup when it is hot, children now blow the soup by themselves. To cool hot chocolate, they take the milk and pour it in by themselves instead of expecting someone else to do it.

Even though children begin with such *small steps of production*, their understanding has already reached an impressive extent. They now take part in complex events and anticipate longer sequences of activities from their surroundings.

It will always be like this: The understanding will be more comprehensive than the anticipation; the anticipation will be more comprehensive than the production.

Children begin to produce some steps of problem solving events in daily living *when the situation for those steps is directly given*, e. g., a child sees a tomato and a knife on the table, takes the knife - often to the horror of those who are around - and tries to cut the tomato; or a child sees a shoe and tries to put it on. Beyond the performances of recognition and anticipation, there is now an additional one - a plan. This means that the interiorization of experiences in problem solving events has progressed to such an extent that the child can do more than merely anticipate them. Anticipation must be there, but it is now so comprehensive that, given the corresponding situation, the child can now produce the anticipated movement without assistance.

The following example of a normal child illustrates this concept:

Today is E.'s third birthday.

What a huge birthday cake!
Can he blow out the candles by himself?
Yes.

Can he cut the pieces of cake by himself? No.
His godfather guides his hands. He can now touch, embrace, move, and feel the changes.

Here is a piece of cake. "Mmm"! Now he can eat it!

At first, when children and adults who are perceptually disordered begin to produce some activities spontaneously, those around them are quite happy. They have been waiting for this moment for a long time. They have guided with patience and perseverance. However, they are greatly disappointed when peculiarities soon appear which are not understandable.
In the following paragraph we will try to describe this development:

1.2 Initial Acts of Production Are Eagerly Awaited – But?

Usually children and adults who are perceptually disordered begin with brief activities which almost exclusively include objects within their field of action.
At this time their behavior can change quite rapidly. Up to this point you could guide them easily. Now they begin to wriggle away from being guided. Previously they would only watch what was going on around them; now they start to set things in motion. Their quiet behavior becomes hectic.

A., 8 years and perceptually disordered, is presented to a group of therapists within a teaching situation.

It is reported that, although he has been guided for two years, he seems to be regressing. His therapist also reports that he is barely tolerable any longer.

What is happening?

A video recording made a year earlier was shown of him in a situation where he was being guided. He appeared attentive throughout the event. The therapist was quite satisfied! And now? Any attempt to do the same thing fails. He shows hectic behavior all through a guided event. Even the therapist who is trying to guide him becomes tense.

Is this regression? How is this kind of behavior elicited? We begin to observe A. in his spontaneous activities. Whatever objects he sees, he runs to them, touches them, grasps them, and sets them in motion. First this happens at a normal speed, but soon his speed increases. Finally, the child is tense, and the surroundings are a mess.
An analysis of these observations leads us to the following conclusions: A year before, A. appeared to be hardly aware of his surroundings. He didn't touch things, and consequently, the objects around him were always in their places. He never changed anything. But today? He sees something, runs to it, touches it, and takes it off. We interpret: he now knows that what he sees, he can also touch - it exists. Even more, he expects that what he touches he can move and take off. He produces these steps. Thus he has learned that he can act upon - the first sign that he begins to interiorize causes and effects of the Wirklichkeit. His new kind of behavior is an indication of progress, not of regression. But what should those around him be doing? How should they encounter his hectic behavior?
Before we discuss this problem more thoroughly, we will report a corresponding observation involving an adult.

*Mr. Z. is brain damaged and receives therapy at a clinic. For several weeks Mr. Z. could be guided without difficulty to solve problems in daily living situations. He was attentive. On the ward, he was considered to be a quiet patient. He usually just sat in the hallway and looked around. He bothered no one. If someone asked him to help carry something, like the basket of wash, he was quite willing.
The situation changed rather suddenly, though.
Now, in therapy he can hardly be guided. He begins to talk, makes faces, and rubs his hands over his face, tightly closing his eyelids. On the ward, he walks around and continually opens doors, asking anyone he meets to bring him a glass of water (even if he has just had one).
At the clinic, they begin to discuss his regression and to question whether it would be meaningful for him to remain in their care.*

We began to collect observations and analyze video recordings which showed him being guided during various events. We also observed the therapists working with him.
Using this procedure, we reached the conclusion that Mr. Z. had begun his first productions. Now, when he sees objects, he doesn't expect someone else to do something with them. He can perform the movements himself. Consequently, when he notices a closed door, he can open it by himself; when he sees cards lying on a table, he can grasp them. But, he cannot go beyond that. He cannot yet solve problems of daily living. They are still too complex and he is conscious of his failure. It makes him tense. His tension is expressed through facial contortions and tightly closed eyelids or by rubbing his hand over his face. After considering all his actions, we conclude that his behavior expresses progression rather than regression!
Arriving at this conclusion, we must ask some questions. What are the characteristics of these productions? Mr. Z. opens all the doors within his field of action and grasps all the cards he can reach. The child, A., sets in motion every object he can touch. These kinds of productions demand little planning

– only a simple kind of decision, selection, or adaptation. They can be performed in a mechanical and monotonous way. They become *habits*. Mr. Z. has the habit of opening doors and A. has the habit of setting everything in motion. Shouldn't we be happy that the two of them have reached the level of habit production?

We need to discuss the concept of habit formation. What are the positive aspects of habit formation? What are the negative ones? When should we be glad that those who are perceptually disordered have habits? When do habits become obstructive?

1.3 It Is Good to Have Habits

We all learn habits. As soon as we are in a new environment we can observe how we acquire habits.

Quickly I begin to take the same turns when going to work. While eating, I sit on the same chair. I return the book I'm reading to the same place. So, what makes the formation of habits so special?

Habits unburden me. When I always put the book back in the same place, I am saving time and energy. I do not have to decide each time where I should put it. When I sit down to read, I can find it faster. When I take the same route to work each day, I can walk without having to think of which turn to take or where to go now. I do not have to look around continually. I recognize the houses along my way. I do not have to look at them carefully. Instead, I can think about my preparation for the day, or I can consider what happened yesterday.

How exhausted I would be if I didn't have so many habits. How could I perform if I had to focus my full attention on every activity? How would it be if typing were not a habit, and I had to search for every letter by looking at each one every time? Or when I consider my work in the kitchen, what would it be like if I had to concentrate fully on how to cut a potato or how to peel a carrot or how to wash the dishes! Instead, I use my habit of how to hold a knife when cutting or peeling. I do not have to think of the "how" – it happens automatically. This word – automatically – helps us to characterize a habit. Only parts of our brains have to be active when performing a habit. We can use the other parts for other problems. Thus, I can enjoy knitting and reading at the same time.

Habits are a great help in daily life. They are even necessary for survival, but they are helpful only as long as we know when and where they are usable.

In order to know the usefulness of habits, we must be conscious of their restrictions.

1.4 Habits Do Not Make Progress Happen

Two features of habits explain why they do not make one progress. One is related to the direction a habit takes; the other one is related to the situation in which the habit is performed.

1.4.1 Habits Are Rigid

Piaget (1950) described the characteristics of habits. Habits invariably follow the same direction. When I look up a word in the dictionary, I repeat the alphabet. This always occurs in the same direction – a, b, c,... I can hardly reverse the direction. To do so, I would have to learn another habit. Thus, the *sequential steps* of habitual activities are *rigid*.

The following example is taken from Katz (1948):

While a caterpillar is in a paralyzed state, the wasp takes it into a hole in the ground already prepared for an egg. Before bringing in the caterpillar, the wasp usually checks the hole. At this time, if someone were to move the caterpillar a short distance away from the hole, the wasp would follow a predictable pattern: Returning to where it originally left the caterpillar and not finding it there, it would begin to look for it. If, after a short search, the wasp would discover the caterpillar, it would again carry it back to the entrance of the hole. One would assume the wasp would now put the caterpillar right into the hole since it has just been checked, but this is not what happens. (pp. 140-141, translated from German).

Katz continues to describe how the wasp has to check the hole each time the caterpillar is removed from the place where it was left – once, twice – as often as the caterpillar is moved.

It is said that we are "slaves to our habits". This means that people expect the events of daily living situations to consist of habits with very little change. They have difficulty accepting an unexpected change in a situation. It is difficult to live with this kind of person for any length of time. They only understand a limited number of circumstances and behave similarly to the wasp. When new problems occur, they cannot change the sequence of their habits and will try to start over and over again with their customary sequence of activities.

1.4.2 Habits Break Down When Situations Change

D., 10 years and perceptually disordered, lives in a residential school. He comes to our Center once a year for an evaluation.

At this time his school teachers report that he is making good progress, but they complain that his parents are unable to teach him discipline. This is most apparent when he travels home with his mother on Friday nights. He does not obey her and walks all over the train. At home, there is chaos, as he will not be quiet for one minute, although at school he can sit down quietly for long periods of time.

It appears that D. has begun to produce some daily habits. He can perform them in a familiar surrounding – in the school and in the therapy room – but when he is in another surrounding, these performances break down.

Evaluations like this one are frequent for us. What is at the bottom of them? When I receive this kind of report, I try to evaluate both the school and therapy situations. Is the child in the same environment day after day? In the same classroom? The same therapy room? Perhaps sitting at the same table and on the same chair? Or has it been considered to take him or her on a public bus? Or to travel to town? Or to go shopping in a self-service store? Or to walk into the middle of a group of busy shoppers? Has the child been placed in unfamiliar situations?

There is always a danger that children and adults who are perceptually disordered learn certain performances in the confines of highly protected areas, e.g., the home, therapy rooms, classrooms, and clinical situations. Their performances, therefore, become habits. When they leave a highly protected environment, they meet changing situations and their habit performances break down (see Part II, paragraph 2.2.6 "...Then the Wirklichkeit Slips Away").

Most of the time, we do not realize to what degree situations repeat themselves in familiar surroundings. Some examples from a Swiss school for perceptually disordered children follows:

There had been a celebration at the school the day before, and the kitchen, which had been used for the occasion, was not in order yet; furniture was still stored there.

The children usually eat at a huge table in this kitchen, but today they will have to eat in the hallway next to the kitchen.

Cheese and potatoes are ready on a counter in the kitchen so that the children can serve themselves. Plates and forks and knives are ready on a table in the hallway.

Everyone sits down, sings the meal song, and wishes each other, "Bon appetite". This is the usual ritual.

The children jabber about being hungry. One of the teachers takes her plate and goes into the kitchen to help herself to cheese and potatoes. She comes back with her plate full and begins to eat.

The children repeat that they are hungry. S. shouts that she will go to the kitchen and get potatoes. However, she doesn't stand up to do it. R. takes his plate, goes into the kitchen, but returns with it empty. K. is guided by the teacher but her body is stiff from tension. What an effort it is to master this kind of unfamiliar situation!

Reports indicated that K. had been able to prepare coffee in the classroom.
When a visitor arrives, K. is asked to prepare coffee for the guest, but his habitual performance breaks down in this new situation.

According to his mother, P. has quite often squeezed oranges for juice. Therefore, preparing orange juice should not present any problem for him.
In a teaching situation, P. is put in unfamiliar rooms with his grandmother who has never guided him. We try to teach her how to do it, but P.'s habit performance breaks down completely.

There are many more similar examples.
When we think about them, we are reminded of the discussion about competence and performance at the end of Part II (see paragraph 3.3.4 "...and the Competence Does Not Become Performance"). We should be pleased when a child finally sits quietly for a period of time in the therapy room because this kind of behavior is now in the child's competence. Being competent does not mean that the child can perform in every situation.
We seldom think about the importance of the concept of changing situations. Day by day, we have the same routines and situations which are quite similar. We are dominated by habits and can hardly imagine what it means when the complexity of a situation increases so much that an insufficient amount of information is received, and thus performance breaks down.

We are reminded of the difficulties in transferring a performance learned in one situation to another one encountered by adults who are perceptually disordered. In the therapy room, these patients learn how to walk, but once they are outside, they are unable to walk (see Part II, paragraph 2.2.6 "...Then the Wirklichkeit Slips Away").

If you would recall the examples there of the perceptually disordered children and adults who were beginning to produce their first actions, at that time we raised the question: How can one work with those children and adults so that they progress beyond this level?

We will try to answer this question in the next paragraph.

1.5 The Gift of Curiosity – How to Break Through Habit Formation

Piaget (1950) contrasted habit formation with operation. Operation is characterized by mobility and expressed by the possibility of making detours, reversing sequences of activities, and including new ones. Operative handling is characteristic of human behavior and is the basis of human concept formation (Affolter, 1968).

Each spring and fall, herds of sheep graze around the mountain villages of Switzerland. Lambs are born and are soon grazing with the herd. When I watch them, I am fascinated by the behavioral differences between the old sheep and the young ones. The old sheep graze quietly. In an easy way they move on, lifting their heads from time to time to make sure there is no danger, and continue to graze. This is not so with the young ones. They are almost always moving around, except when asleep. They turn their heads toward any noise, scent the different smells, jump at one place – usually several times – they look around, listen, smell, and get to know the world around them. Their whole behavior expresses *curiosity*. It is so different from the behavior of the older sheep!

Curiosity – what does that mean? Why do I call the lambs curious and use this quality to differentiate between the old and young sheep? How can I describe that difference?

When I observe the older sheep, they hardly ever lift up their heads. Most of the time, their attention is directed toward the grass. Only rarely do they look around to check what the young ones are doing or to see if there is any danger. They graze and graze and then lie down to chew their cud, closing their eyes in a sleepy way. The behavior of the young ones contrasts sharply: Their legs and heads are always in motion. They are *continually looking for new events* – being aware of any little change in the situation and approaching any unfamiliar circumstances. They are always ready to withdraw as soon as the unknown moves too rapidly, but shortly afterward, they begin again to search for unfamiliar events and never tire of being eager for new kinds of information.

Receiving new facts, transmitting information, changing situations, searching for corresponding information – again and again – these are the means of preventing habit formation or breaking through habits which are already formed.

From the example of the sheep, we can make an important conclusion: Curiosity is obviously something important for getting to know about the Wirklichkeit, and ultimately, learning and developing.

Our perceptually disordered children and adults have severe difficulty in satisfying their eagerness for new kinds of information. The reason is because of their pronounced *poverty of information* due to their disordered perception.

1.5.1 If Information Is Restricted, How Can It Be Expanded?

In discussing the behavioral peculiarities of children and adults who are perceptually disordered, we pointed out that their *productions are characterized by a poverty of information* (see Part II, Chapter 3, "What Happens When There Is a Lack of Tactile-Kinesthetic Information"?).

The information that children and adults on the level of first production are able to take out of a situation is extremely restricted. They take the closest object and perform an activity with it. Similar kinds of behavior can also be observed in ourselves. When I am tired and have chocolate nearby, I will often eat it until nothing is left. In a stressful situation I, too, am restricted in the amount of information I can receive. I consider only what is next to me, and I eat all the chocolate.

Since this is an important phenomenon, we offer another example.

Ch., 23 years, with a head injury, is part of a teaching situation. The participants had guided him in preparing "sausages-in-crusts".
There are 14 people sitting around the table. Ch. distributes all the sausages. His therapist is sitting next to him and does not get one.
The therapist says, "Look Ch., I don't have a sausage". Ch. looks at the therapist's plate; his face shows that he feels sorry. He says, "You can have some of mine".
Ch. takes a knife and fork, cuts the sausage in half, picks up one half with his fork, and puts it into his own mouth.

Some would say that this was selfishness. Is this true? What has really happened?

Two days later the participants prepare a cake with Ch. Everyone receives a piece and two pieces are left. Ch. says, "I will take the pieces to B. and L. (his therapists at the clinic)".
Ch. has to wait for a taxi. He sits next to a table where the two pieces of cake are on a

251

plate. Ch. is talking with two of the course participants. What happens? During his conversation, he takes one bite after another until there is nothing left on the plate. When the taxi comes, the cake has disappeared.

Selfishness again? Or, what has happened?

Ch. perceived the situation visually. His neighbor had no sausage on his plate. Ch. could verbalize the visual information. This is visual recognition. He even expressed a solution, "You can have some of mine". The anticipation led to an activity, a first production. Ch. cut the sausage in half. Then came the difficulty. Ch. could cut a sausage and put it into a mouth – *his* mouth. This is the habitual sequence – the sequence which requires the least amount of information in order to be performed. Ch. could not master the complexity of the correct solution: He could not put the sausage on the therapist's plate so that the therapist could eat it rather than himself. In other words, he could not perform the required *planning of a detour*.

Ch.'s difficulty is one of the behavioral peculiarities which can be observed in children and adults who are perceptually disordered and who are restricted in their amount of tactile-kinesthetic information.

Productions soon have the characteristics of habits, and performances lag behind when a detour is required. Perceptually disordered children and adults repeat newly learned activities in a rigid form as long as they are at this first level of production.

The following examples are of perceptually disordered children who are at this first level of production:

K., 10 years, is at the school camp.
In the morning, as she helps to do the laundry, it is time to open a new box of detergent. She is guided in cutting it open with a pair of scissors.
The next day she helps with the laundry again. She takes the open box of detergent and cuts off a piece of the box, just like she did the day before.

Another time, she is supposed to put a dish of food into the oven. She puts on the oven mitts even though the oven and dish are cold.

D., 13 years, helps to prepare prunes in the kitchen. He is being guided. With a prune in the left hand and the knife in the right one, they cut the prunes into halves, take out the stones, and put the halved fruit into a bowl. D. wants to continue by himself. Later, someone tries to show him another way to take the stones out. He balks and says, "No. That's not the way".

T., 11 years, puts a piece of cucumber on a buttered slice of bread. His mother reports that ever since that activity, he won't eat a slice of bread unless there is a cucumber on it.

At this level, productions are learned without considering changes in a situation. A piece of cucumber has to be on a slice of bread whether it is eaten at school or at home. The child puts on oven mitts even there is no heat in the oven. Likewise, when opening a bag of laundry soap, the child cuts a piece from it without considering whether it is already open. *Learning on this level includes highly restricted information.*

The examples of 8-year-old A., Mr. Z., 10-year-old D., and 23-year-old Ch. indicate that perceptually disordered persons who are on the level of first productions or habits, are often misjudged by those around them in the environment. They are called restless, hyperactive, tyrannic, selfish, etc. Usually, when patients are referred to us it is because they appear intolerable to those around them due to their behavior or apparent regression. They are probably functioning on the level of habit formation.

When we reinforce the formation of habits, we do not allow these children and adults to progress in development. They become like machines, turning and producing as long as the environment does not change. Development is blocked, though, and the moment comes when the performance of the "machine" breaks down. At these times, those in the environment usually refer to them as being poorly motivated.

This interpretation allows those in the environment to place the blame for the performance breakdown on the disordered child or adult. This could mean that the educators and therapists do not necessarily have to think about what is occurring during the learning situations nor possibly conclude that a change in their procedure is indicated.

Sometimes, instead of referring to it as poor motivation, this behavior is attributed to "changes in personality". This was the case with Ch. described in the situation with the sausage. At the clinic he attended, there were few disordered patients. Therefore, situations changed quite often and became too complex for him. He panicked frequently and his performances broke down. One day it was concluded that he would have to go to a psychiatric clinic. The guided kind of therapy he had been receiving would have to end because it appeared that he was only regressing and becoming increasingly intolerable. This became a very critical and uncertain period.

When children and adults who are perceptually disordered begin with first productions, we must be extremely careful to transmit adequate tactile-kinesthetic information during guided events. This means that the *creation of changes in resistance* and maximum contrasts when exploring neighboring relationships - taking off and bringing back; touching and moving; and embracing and moving - which include the support, must be perceived over the tactile-kinesthetic sensory system and be connected with problem solving events in daily living where guidance is given again and again.

We tried the preceding ideas with A., who was described on pages 246-247. The dishes were washed and dried. Now A. must put them back in the cupboard. We were at the sink with the dishes but the cupboard was across the room. The dishes belong there but how can one guide a hectic boy like A. across the room? We thought about it. How would small children hold a dish so that it wouldn't slip to the floor?

They would grasp it with both hands and press it very tightly against their bodies, making sure it is in their hands, being firmly touched and held.

So, with the plate pressed tightly against his body, A. is guided across the room to the cupboard. Once there, other changes in resistance are elicited and perceived. Only then are the dishes put on the shelf. As they proceeded, we watched with surprise. The boy who had appeared to be so easily excited became increasingly calmer as the task progressed.

Changes in resistance have to be elicited. To do this, I need *variations in movement.* For instance, after I sit on a chair for a few moments, I do not feel its hardness. Only when I move, do I feel how hard the chair really is. Recall the examples in Part II, paragraph 2.1.2 "They Know About the Rules of the Stable Support and the Side".

A., 9 years and perceptually disordered, is scooping out the pulpy part of a lemon. She is being guided.

We observe when she uses her right hand or her left, when she uses her hands or a tool. We also observe the changes in resistance she is eliciting between the support, the side, and her own body.

A. uses the spoon with her body functioning as the stable side. While the right hand holds the lemon, the left one performs the movement.

Both hands touch, embrace, and move while the table offers the resistance of a support.

With the table as the stable support, the left hand moves the spoon while the right one holds the lemon.

Now the hands change. The left one takes over the touching and taking off the pulp while the right one holds the spoon.

1.5.2 Feeling Causes and Effects

Another aspect of transmitting important information is the *exploring of cause-effect sequences* enclosed in the problem solving activities of daily living. It should be self-evident that this transmission of information happens through feeling the changes in resistance.

A few examples will illustrate this:

Recall the example of K. who cut open the box of detergent even though it had been cut open the day before (p. 252). She had learned that she has to take the scissors and cut the box. She didn't notice that cutting causes the box to be open and consequently the detergent to come out.

When it becomes clear to the teacher that K. did not get these cause-effect relationships, she changes the sequence of solving the problem.

She provides another closed box of detergent and guides K.'s hand in taking it and holding it over the basin containing some dirty clothes and water. Nothing happens! No detergent falls into the water; the box stays full. There are no changes.

Now, she again guides K. with the scissors and cuts open the box. Instantly detergent falls out of the box into the water – cutting – and its effects – detergent coming out.

M., 8 years, refuses to take off his clothes when it is time to get into the bathtub. One day the teacher allows him go into the tub with his clothes on. What a surprise to him! M. immediately gets out of the water with his sopping wet clothes. The next day there is no problem for M.; he removes his clothes to take a bath.

It is possible to provide those who are perceptually disordered with more tactile-kinesthetic information about the connection between causes and effects.

We should approach the problem of using tools in a similar manner. As was mentioned previously, tools are taken when a manipulation by hand is not successful; thus, we should first try to perform the causative action *for solving a problem with the hand.*

B., 9 years and perceptually disordered, tries to divide an apple. He is being guided.
We hold the apple firmly on the support. With one hand, we remove the stem first. Then we try to divide the apple in half – by hand. We press several times without success. Even though our fingernails break through the peel into the pulp of the apple, there is still no maximum change in resistance. After making another attempt, we get the knife and cut through the apple.

With Z., 9 years and perceptually disordered, we try to crack open a nut. We press the nut on the support but there is no change. We squeeze the nut in one hand and press it with the other one. Still there is no change. Finally, we take a stone and pound on the nut; it breaks!

By embedding the person's own body into the exploration of cause-effect relationships, we reach another goal. In addition to information about the causes of changes in resistance, *we provide information about the quality of the surroundings.* This kind of information is important for getting familiar with the surroundings (see Part I A, paragraph 1.4 "The Surroundings Become Familiar"). The information a person could receive about the qualities of an apple would be quite poor if, from the beginning, I used the knife to cut the apple. However, a tremendous amount of information is received when I try to break the apple with my fingers. By doing this, I can learn much about the characteristics of an apple – a basic experience for concept formation about an apple.

To Summarize:

In order to break through the formation of habits, we must transmit information about changes in resistance within the guided problem solving events – changes in resistance when touching, embracing, moving, releasing – among support, objects, and the body. Again and again, we apply the rules of touching – the support and the sides – as well as the rules of acting upon, taking off, and the neighborhood.

With such a restricted set of rules, we can lay hold of an indefinite number of situations about causes and effects which will benefit the children and adults who are perceptually disordered!

We tried this with A., with Mr. Z., and also with Ch. For the longest time, we could follow the progress of Ch. We mentioned earlier (see p. 253) that the necessity of referring Ch. to a psychiatric clinic was discussed due to his panic reactions being interpreted by those in his environment as aggressive behavior. The referral needed to be reversed. We attempted explanations of Ch.'s deviant reactions to the environment – again and again. In therapy, we tried to break through his habits, to feel the changes of situations with him, and to enlarge his understanding of problem solving events. Little by little, his panic reactions decreased, and his behavior adapted better to changes in situations. Today, a few years later, he lives independently and works full time.

1.6 The Way Back to the Main Road Is Found by Using Detours

I continue to work with children and adults who are perceptually disordered, guiding them through problem solving events and providing them with information about changes in resistance – guided and felt! Habits will be formed, but I will break through them. Curiosity will appear and understanding will expand – but when will the production of actions become more flexible? Depending on the degree and the kind of perceptual disorder, this will occur early for some but later for others.

A first sign of attaining more flexibility is the *behavior of detours.* An example of a normal child illustrates this ability to change:

M., 18 months, knows how to open doors. When she sees one ajar, she pushes it all the way open.

A few days ago, she also discovered that if she manipulates the doorknob, the door would open even when she doesn't see a gap. She also notices that not all doors can be opened by turning the doorknob and pushing the door.

We know this to be the case when the door can only be opened by turning the doorknob and pulling backward on it. Even though our bodies move forward through the door, the door must first be moved in the opposite direction. Such a situation requires the understanding of detours. M. had not yet reached this level of production.

During the second year of growth, we can observe the appearance of this kind of detour behavior. This behavior increases in frequency, and more and more detours are included in the daily events of solving problems. For instance, when children see something they want in a cupboard, but are too short, they go for a chair and climb on it to reach their goal. M. learns now to open also those doors which require a detour: Turn the doorknob, pull the door backward – even though she must eventually go forward through the door – and then pass through the opening, closing the door again.

Being able to use detours is the expression of an increase in the interiorization of experiences for solving problems in daily living. I can go and get a chair as a detour because I can interiorally represent what will happen next. M. can pull the door backward because she can represent the next steps:

The door opens, and she squeezes through the opening; she pulls the door towards herself until it meets resistance; and finally, she closes the door again.

With the possibility of planning detours, children's activities become more flexible. Detours can also be used to explore the surroundings, as the following example of a perceptually disordered child illustrates:

For many years, T., 14 years, had formed habits in a very rigid way. For instance, as soon as he would notice a round object near him, he would grasp it and skillfully rotate it. Slowly he began to show more and more understanding for problem solving events. This was followed by initial productions in daily living situations.

We continued to guide him during problem solving events, providing information over and over again.

Finally it happened! Detour behavior was observable when he solved problems in daily living, and he even began to use detours when exploring his acting upon the environment. Here is an example:

T. is on a hike with the school. The road is stony. T. takes two stones off the road and carefully puts them on his shoes. "Will they stay there"?

"Can I still walk this way"?
Carefully – slowly – T. puts one foot on the ground and lifts the other one.

The *use of resources* is closely related to the behavior of detours. Among the resources, tools are of special interest to children.

E., 3 years and a normal child, is in his father's workshop. What a magnificent field for action!

E. scratches on a wooden board and discovers a hole where he could put a nail. ▶

"Does the nail fit into the hole"?

"It does"! Now he uses the hammer. The nail goes into the hole.

"Here is another nail – not firmly in a hole".
E. hammers that nail, also.

Now the nail is fastened firmly in the support.
(It is bent but this does not bother E.)

"Here is another nail to hammer"! ▶

"Now I'm done"! (I've mastered all of these problems.)

"I better go and put the hammer back where it belongs"!

Besides detour behavior and keen interest in resources, the *curiosity about cause-effect relationships* is a sign of growing flexibility of production.

T., 14 years and perceptually disordered, was previously described in a situation with stones (p. 257). Now he is grating nutmeg and cheese to prepare a toasted cheese sandwich. He is working by himself.
First, he grates the nutmeg.

▶

He rubs the nut over the grater. He observes the effect. "Will this make nutmeg powder"?

▲
"How does it taste"?
He licks the rough surface of the grater with his tongue.

◄
Now he will grate the cheese.
He tries to open the plastic with his teeth.

▼
Is he successful?
He examines the plastic. "Does it have a hole in it? Yes, here".

He takes off the rind and holds the cheese in his hand.

"Can one grate it"?
He tries, and at the same time, he looks through the opening. "Is it possible to see the grated cheese coming out of the little holes of the grater"?

"No. There is no cheese to see. Where is it"? There, sticking to the grater. He scratches the cheese from the outside surface of the grater with his finger...

...and then scratches from the inside.

In this way, T.'s behavior in living and dealing with daily problems becomes more and more flexible – adapted. When difficulties arise, he doesn't become panicky as quickly as before. He can now explore some problem situations, pick up information about causes and effects, and change his performance accordingly.

Here is another example about T:

T. attempts to tie a bundle of sticks together, i.e., make a fagot.
How can he make the bundle stay together? What detours does he have to consider?

What are his resources? String? Wire? Mouth? What are the causes and effects? Do the detours work out?

T. is now able to plan and perform detours. In addition, he can find his way back to the main goal. In our example, he adapts the effects of string and wire by exploring detours and orienting them toward the main goal of making a fagot. He finds his way back to the main road even after producing detours.

The following two examples further describe T:

An egg shell has been broken while boiling.

T. licks the egg. ▶

▲
He looks at what has happened.

He grasps the egg white with two fingers – the tweezers grasp.

▶
With just those two fingers, he removes the egg white...

▼
...and lets it fall into a bowl. He only uses one hand.

He looks inside to see whether the eggshell is empty...

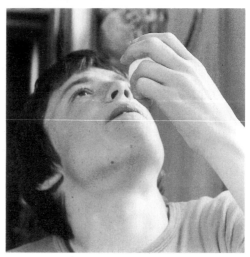

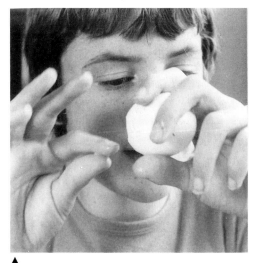

▲
...and licks it one more time. His free hand is in a "grasping position".

T. licks his finger and examines the egg again. Note that the fingers of the left hand are seizing the egg instead of embracing it!
▶

Some egg white is still in the egg. He tries to make it fall into his mouth. (He doesn't know how to suck it out.)

T. wants to eat an ice cream cone.   ▼
He opens the wrapper.

He touches the cone.
Note again that he uses a seizing pattern.

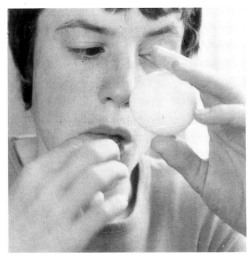

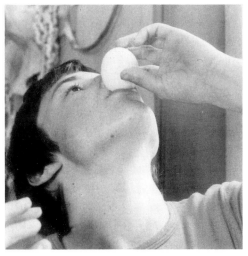

He begins to lick it.

"How else can one lick it"?
T. examines the situation.

"Here is the top; ▼

▲
...here is the side;

◄
...and now the whole thing is on the table".

The resistance gives way.
Again, note the seizing pattern, and consequently, a lack of information that may have been gained by embracing.

To Summarize:

In this chapter "Production Begins" we discussed the progress of children and adults who are perceptually disordered. We indicated that, because of a lack of adequate information, their productions rapidly take on the characteristics of habits.

Habits can be of tremendous help in daily living. However, they do not cause those who are perceptually disordered to progress in their development.

In order to progress, they need information – tactile-kinesthetically perceived changes in resistance within problem solving events in daily living.

The more expanded the interiorized experiences in problem solving events are, the better the productions. As the productions become more flexible and more adaptive, detours can be embedded.

Still, those who are perceptually disordered will continue to need the transmission of tactile-kinesthetic information in the form of guided problem solving events in daily living. Only when they receive this will they show further progress in understanding and exhibit new and different kinds of productions. Beyond understanding and production, comes yet another important performance – representation. This last performance includes language. We will discuss this progress in the next chapter.

2 Return to the Problem Solving Events of Daily Living – Then Comes Representation

Interiorization of tactile-kinesthetic experiences with problem solving events in daily living is not only necessary to reach a level of anticipation and, ultimately, production, but it is also necessary to reach the level of *semiotic performance*.

A semiotic performance consists of taking any kind of form to express a content referring to past or to future events. Piaget (1962) pointed out the differentiation between a form and its content – between the designator and the designated – as important criterion for semiotic performance. An example of this is a child who takes a small stone off the ground, hides it in his or her hand, and asks, "Do you know what I have in my hand"? He or she immediately responds, "A candy". Opening the hand, he or she says, "You know, it's not a real candy. You can't put it in your mouth. Still it's a candy".

The child behaves *as if* the stone were a candy. He or she takes the stone as a *form* to represent the *content* of candy. The child behaves this way when there is no candy in the actual situation and even when there is no expectation of receiving any candy either. The child retrieves previous experiences with candies and expresses them with the picking up of a stone. This is called *representation*. It is interesting to note that the stone is taken to represent a variety of contents. Now it represents candy; at another time, it might be a loaf of bread.

Various forms can be used to express the content of past or future events. We use forms that can be seen (optical), for instance, graphic forms of pictures or letters. The most frequent ones are forms we hear (acoustic), e.g., speech sounds. There are also forms we can feel, such as, Braille which is used to communicate with blind people. Some forms are perceived through several sensory systems. When I speak, I perceive the speech sounds I am producing over my tactile-kinesthetic *and* auditory systems. When a child does not learn adequate speech sound production, the cause may be due to a disorder in the tactile-kinesthetic system or to a hearing loss. (There are also other factors to consider, such as, anomalies in the structure of speech organs, etc.)

Forms can present some similarity to the content they represent. These are called *symbols* (Piaget, 1962). The child in our example used a stone to represent candy because it has a certain resemblance to candy.

This performance is of a symbolic kind. The child selected that form and I can guess its content. Such individual symbols are also utilized by deaf children.

Language is different because the forms of language do not resemble their content (the only exceptions are onomatopoeic words which imitate the noise some objects may emit). When I listen to people speaking Chinese, I cannot guess what they are talking about because the sounds do not resemble what they represent. They have been chosen *arbitrarily*. Verbal forms are also *conventional*. This means they have been determined by social groups on a conventional base (whereby the historical dimension is omitted). Piaget (1962) called them *signs*.

Children discover language around 18 months of age. This occurs at the same time as the discovery of symbols. Both language behavior and nonlinguistic symbolic behavior are different expressions of a "representative function" – semiotic performance (see Piaget). In detail, Piaget (1962) described the preceding levels and the discovery of semiotic performance, such as, symbolic behavior and language. The first level is understanding/comprehension, just as it is for other performances; production comes later.

In the following paragraphs, we will again focus on the development of normal children. First, we will observe their symbolic behavior; then, we will discuss their discovery and their initial handling of language. The observations of normal children will be contrasted with some observations of perceptually disordered children and adults.

The reader is reminded that the discussion is restricted to the earliest level of *symbolic behavior and language*. Discussion about further development of language is beyond the scope of this book.

2.1 The Symbol Represents Problem Solving Events of Daily Living

Problem solving events and the development of perceptual performances are now interiorized to such an extent that children discover something new. They begin to represent problem solving events which occur frequently in their surroundings by utilizing symbolic forms. Such forms can be found ready-made in their environment, or the children can create them.

2.1.1 Problem Solving Events Are Represented by Forms Which Are Ready-Made

J., 2 years, 6 months, spends some time in the mountains. One of the neighbors has a stable for sheep. The sheep graze around the house where J. is staying. When she goes for walks, she stops and observes the sheep – especially the lambs. She visits the sheep in the stable. She is allowed to pet the lambs and take them in her arms.

Now, when she finds stones on the ground, she picks them up and begins to play with them. The stones become the sheep.

To make this kind of a representation possible, the forms do not have to be constructed. The stones were there and ready-made. To symbolize the grazing of the sheep, she had only to move the stones around. Similar behavior occurs in the next example:

J., one year later, is in the garden. She explores the different objects around her. Suddenly she remembers that her baby doll is hungry.

She takes her baby into her arms. A little box she has just found serves as a bottle.

Is the baby drinking her bottle?

J., 2 years, 6 months, has a baby sister. She often helps her mother take care of the baby.

Now she represents this kind of problem solving event with her baby doll.

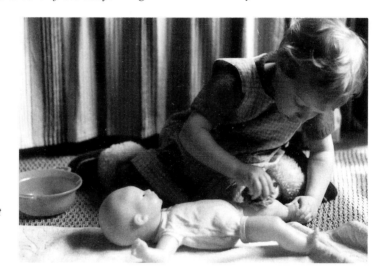

She bathes the baby just like she has seen her mother bathe her sister.

And now – what comes next? Dressing the baby is difficult!

Besides using objects in the environment which are ready-made, children often use their own bodies as a form to represent an event.

After her sister was born, J., 2 years, 6 months, frequently played the role of being a baby herself. She cried and moved in a jerky manner like a baby. Then one had to take her in his or her arms and caress her.

Role playing soon encompasses most of children's play. For a period of time, a child is Petra, the older cousin who can do so many things. At other times, the child is the lady next door who is expecting a baby. After Christmas and for many weeks, the child is Mary.

Often, children symbolize different every day activities with their own bodies. They behave as if they were eating, cooking, setting the table, or going for a walk. All of these activities are represented without objects, i.e., using just their bodies to perform the corresponding movements.

J., 3 years, doesn't sleep as frequently as she used to when riding in a car. As soon as the car stops, though she leans back and pretends that she is asleep. At the same time, I can see a smile on her face.

2.1.2 The Forms for Representation Are Constructed

As the representation of problem solving events becomes more complex and extensive, the forms already in the surroundings are no longer sufficient. New forms need to be constructed. This requires changes in the environment produced by *true problem solving events in order to represent previous experiences in problem solving*.

The following example illustrates this complexity:

J., 2 years, 6 months, wants a stable for her sheep (the stones). They should have one, just like the real sheep do, because it is evening. But, where is the stable?

J. puts one small wooden board next to another one, leaving an opening in front so the sheep can enter the stable. Have all of the sheep found a place in the stable?

Now the sheep are inside. They are ready to sleep for the night, so J. can close the door.

It is interesting to observe how a child at that age constructs spontaneously. J. constructs a barn by putting one wooden board next to another so they touch each other. She applies her touching rules – the rule of the support and the rule of the sides. The information she uses is received through changes in resistance between the support and the sides. An adult watching J.'s activities tries to please her by constructing a nice barn with a roof. It is so big, she can even sit inside. However, J. goes in just once. It seems that she cannot appreciate it. Instead, several more times, she constructs her own model. Maybe her model offers her more information about the different changes in resistance than does the big barn with a roof she can not even touch.

Another example confirms the importance of touching both the support and the side resistances.

After work, it is time to relax – just like the neighbor, an older man, does.
J. finds a board which serves as her bench and sits down. Ahh! Working makes one hungry!

J., 3 years, 6 months, plays riding on a train. First, she constructs the train.

A long wooden board serves as a support. Now the side walls are put in place.

"Can I sit in between"?

Her sister boards the train and sits down. Tickets please!

2.1.3 The Symbol Serves to Explore New Situations

With a growing number of tactile-kinesthetic experiences in problem solving events in daily living, children use an increasingly larger variety of forms. Their explorations of new situations become richer but continue to consist of touching, embracing, and moving. Only now, the children are able to integrate them into more complex events. In this way problem solving events are becoming embedded into other problem solving events. Children retrieve stored events and represent them with the help of forms which are taken from an actual situation. The purpose is to explore a new situation and to become familiar with it. The following example illustrates this:

J., 3 years, 8 months, is hiking in the mountains. The big backpack impresses her and she wants to carry it. However, it is much too heavy. What can she do?

"Now I have a dog. Come, Lady". (A family J. knows has a dog by this name.)

"My dog and I - we go up the mountain".

"Lady is tired. Come, I will help you".

"Gosh! The mountain is steep, and Lady is so heavy".

2.1.4 The Picture Is a Symbol

Toward the end of the first year, children begin to pay attention to objects represented by pictures.

G., 10 months, a normal child, looks outside over and over again at the meadow where cows graze. This happens each time he is on the balcony, and so we ask him, Where is the cow? If we are in the kitchen, where a calendar with a cow painted on it hangs on the wall and we ask him, Where is the cow? he looks at the picture; he reacts the same way.

As shown by this example, children can recognize the visual shape of an object long before they reach the semiotic level. At this prelinguistic level, they appear to perceive objects in pictures the same way they see real objects. Often we can observe them scratching the picture, possibly trying to take off the object represented on it. They relate the content of these visual forms to an actual situation, but not to a past or future event. The cows are present now. This recognition of objects represented by visual forms is, therefore, a much more elementary performance than the semiotic one. Animals can learn to recognize visual shapes, but they do not reach semiotic performance. Chimpanzees can learn sequences of visual shapes as a means to communicate. However, we cannot use the shapes to transmit information to the chimpanzee about an event of the past or of the future. For the chimpanzee, *the content of these forms always refers to some event that will happen right away or is already happening.* For instance, we can tell the chimpanzee that it will get a banana (Terrace, 1979), but getting the banana has to happen now. If I tell the chimpanzee at lunch that it will get a banana later, i. e., when it is dark, the chimpanzee will not understand. Piaget (1962) used the concept of signal to define these kinds of performances which relate an auditory or visual form to a content inherent in an actual situation.

It is important to clearly differentiate between a signal and a semiotic performance. Often, therapists try to make children who are perceptually disordered match pictures with objects and objects with pictures. They hope that by exercising matching performances, the language of the children will progress. They do not realize that these kinds of exercises include only the matching of forms without a reference situation involving a tactile-kinesthetic experience with a problem solving event of the past, i. e., in our example, a graphic form – a picture – is matched to a visual or felt form – an object (Affolter, 1974b). Thus, this example is a performance of signals and not one of language.

Often, one refers to comprehension of language when a child acts according to a verbal command in a respective situation, but this, too, is actually a signal performance. Comprehension of signals occurs on a prelinguistic level (Affolter, 1968).

Not until normal children are about 18 months of age, do they begin with the recognition of pictures as symbols. The criterion is the recognition of actions in a picture. They now request stories but the story is never finished. And then...? they ask even when we have told the ending.

It is obvious now, that children refer to whole events of the past with pictures. The content consists of problem solving events in daily living which children take out of their stored experiences.

For weeks, J., 20 months, takes the same book – fairy tales – She turns the pages until she finds her picture: a woman rocking a baby in her arms. She usually looks at the picture for a long time before she puts the book back on the shelf.

The picture has become a symbol. The form has some resemblance to the content. Similarly, children take a form, such as, a doll that resembles a human figure, and use it for symbolic play. The content of the picture refers to past events, just like the symbolic play, and not to actual situations. With this behavior, children fulfill the conditions of a semiotic performance.

Still, it takes children a long time before they are able to not only understand graphic symbols, but also to produce them – to draw them. There is a wealth of literature about the development of drawing. Among other researchers, Piaget and Inhelder (1956) thoroughly described the different levels of development.

2.2 Problem Solving Events in Daily Living and Then – Language Performance

Learning how to speak, in the sense that language performance is a representing action, is a very complex development. In the following paragraph, we will discuss some of its aspects. We will again depart from the assumption that tactile-kinesthetic experiences of interactions in the form of problem solving events in daily living are the root of all developmental performances.

2.2.1 Children Relate Verbal Forms to Problem Solving Events They Experience

Language is not learned by imitation. Several researchers investigated the role of imitation in language learning. Palermo (1978) summarized the findings. He stressed that children can only imitate what they can produce spontaneously. They imitate what they have in their repertoire. Thus, they learn the speech sounds from their speaking environment, but not by imitating directly. The process is much more complicated.

From birth on, children hear the speech sounds of their environment. During the first year, they begin to babble. Their first babbled sounds are not directly related to their speaking environment. Children of different linguistic groups produce the same babbling sounds. Even deaf children babble.

During their second year, children increasingly begin to take certain features out of the sounds produced by the environment. All over the world, children first take out those distinctive features which present the greatest visual, auditory, and tactile-kinesthetic contrasts. Thus, they produce the first sequences of sounds like /papa/ and then /mama/ (Jakobson, 1969). From these sequences, the different linguistic groups created the pet names for parents, e. g., /papa/ and /mama/ in German, French, Italian, and Spanish; /abba/ and /ema/ in Hebrew.

It takes several years before children discover all of the distinctive features. Only when they reach 6 or 7 years of age can we expect them to produce all of the speech sounds correctly.

During this discovery period, we can pronounce words which children fail, but they will still say them according to their stage of analysis of distinctive features.

The father of M., 2 years, 6 months, has a beard. He points out, "Schnauz (moustache)".
M. repeats, "Nauz".
Then her father says, "Nauz".
The child disagrees, "No, not nauz! Nauz"!

M. could judge that when her father repeated, "nauz," he said it incorrectly. She could analyze the required distinctive auditory features and notice a difference. Therefore, hearing is necessary, but this sort of example shows that just hearing the sound is not sufficient to correctly reproduce the word. In addition to hearing, this performance also requires tactile-kinesthetic processing.

Production of speech sounds, even when they are spoken correctly, is not yet language. In order for speech sounds to become language, they have to refer to content. Just as the production of speech sounds is complex, even more so is the performance of language.

Children hear the speech forms of their environment. They always hear them in connection with events. We can observe that children take out of these events something which is *meaningful* to them. Quite often, this something is not the same thing that an adult would take out of the event.

Children create their own grammar from what they take out of the events and from the speech sounds produced in the environment. This process is a *creative* one. Children produce sentences that have never been pronounced before by their environment (Chomsky, 1957). This creative performance is in contradiction to the assumption that imitation is the basis for learning language.

The following example illustrates this:

J., 2 years, 6 months, is with her mother when she meets an unfamiliar group of people. Her mother introduces her to everyone, and there are mutual greetings.

The next day, I step out on the balcony early in the morning. J. and her father are already there kneeling on the floor beside a box. A young bird we had found the day before is in the box. J.'s father is feeding it while J. watches everything intensely. As she sees me, she calls out enthusiastically, "Look! Beautiful blue eyes"! I hesitate. The bird doesn't have blue eyes. I wondered why J. said that, but I didn't reply in disagreement.

How can the child's response be explained? As soon as I have a few quiet moments, I ponder the event. I had heard of the previous day's experience when J. met the group of people. She was in the middle of so many people she didn't know, and everyone was looking at her. It is quite possible that someone mentioned her blue eyes, saying, "What beautiful blue eyes she has". It is obvious that J. did not store the meaning that had been intended. For her, the situation was very emotional, a situation in which someone said something nice about her, the stranger. It was likewise with the bird. The bird is J., and J. is the adult who says something nice using that sentence which she heard during the exchange of greetings.

The next example is a similar one. It again emphasizes the creative aspect of learning language. The content and the referring of content to a linguistic form is related to events which children experience differently from adults.

J., 22 months, was on a visit with her parents. During the visit, the following two photographs of her family were taken with all of them seated around the dinner table. Her father is in the foreground.

A few days later, we are looking at the processed photographs.

J. remarks, "Papa swim".

"Papa swim, no".
I am perplexed.

Swimming occurs during the summer, and although J. had gone swimming several times with her parents, it's October now. I look at the photographs again. Why does J. talk about swimming?

I try to recall a situation involving swimming. The family spends most of the time near the children's pool where the water is shallow. Several times, though, J.'s father left the family to go to a larger pool for swimmers, and J. often accompanied him. On those occasions, she must have watched him going down the pool stairs into the deeper water. First, she watched how her father's legs went under the water and disappeared. Then the trunk of his body did the same thing. Finally, she could only see her father's head sticking out of the water. The first photograph shows her father just like this - only her father's head shows. His body seems to have disappeared in the chair, just like when he was swimming in the water. The second photograph shows her father like when he came out of the water and she could see his body again.

In this way children appear to relate spoken forms to events which they experience.

There are verbal forms in which the content has little to do with real events. Names for colors belong to this group. In spite of children being able to discriminate colors at an early age, it takes them a long time to be able to name colors correctly. However, such discrimination is always related to real events.

J., 2 years, 6 months, often helps to set the table. The various members of the family have different colored napkins. J. takes the blue one and declares that it belongs to G. She takes the pink one and says that it belongs to Papa, and then takes the yellow one and says that it belongs to Mamma. She continues in this manner and correctly matches seven napkins to corresponding people. She doesn't make one mistake with the matching of all those colors, and yet, at the age of 4 years, 6 months, she incorrectly named colors. I bought new boots that were bright yellow. When J. discovered them, she exclaimed, "Such beautiful red boots"!

M., 3 years, is to bring some Pampers to his mother from the blue box. He hurries away, and after some time, he comes back, confused and without the Pampers. He asks his moth-

er, "Do I get the sky Pampers or the grass Pampers (meaning: from the blue box or from the green one)"?

Frequently, children take out of an event, and even out of spoken forms, those words which have to do with tactile-kinesthetic information, and not with visual information.

We are riding on an air cable car. Twice we go over a support structure. F., 2 years, 6 months, is with us. Each time the ride is stopped by the support, we explain to her that it is a "pylon". Each time, she repeats to herself, "Shakes".

When children hear a verbal phrase in a song or prayer or other conventional situation, they connect those forms with some content from their own felt circle of experience. This is the case in the following examples:

D., 5 years, asks for the song "Santa Claus with the sled". His mother doesn't understand. She can't think of such a song. D. corrects himself and says, "No, Santa Claus with the Schi". (Schi is the Swiss word for ski, but it also means to shine as a lantern). As his mother thinks about it, she finally realizes that D. wants the song of "Santa Claus with the 'schi'" (in the sense of lantern).

D. probably has a high temperature, as his forehead is hot and his cheeks are red. His mother tells him, "I am going to get the Fiebermesser," (thermometer). D. cries out, horrified, "No! No cutting". (In German, messen/messer means to measure, but Messer also means knife).

In schools children learn to use *graphic signs*, e.g., reading, writing, and arithmetic, as well as verbal ones. The processes are similar to language. In each case, there are forms which represent events of the past or of the future in a semiotic way. The forms do not have a similarity with the content.

The signs are arbitrary and based on conventions. Thus, experiences with problem solving events in daily living also provide the root for learning to read, write, and do arithmetic.

I am visiting with a family.
M., 2 years, 6 months, is sitting on my lap at a table where there is a bowl with some grapes in it. She is busy taking two grapes at a time in her hand. She puts one grape into her own mouth and the other one into mine. She then takes two more grapes and does the same thing. Suddenly, she notices that her older sister is close by. She quickly reaches into the bowl and fills her hand with as many grapes as she can hold.

There are many different aspects to this kind of problem solving event – changing quantities, matching, addition, subtraction, division, etc. Children store these experiences in their minds and bring them to school where they are taught to relate them to conventional forms.

2.2.2 Problem Solving Events in Daily Living Are the Basis for Deep Structure and for Surface Structure

Every language includes a restricted set of rules which comprise its grammar. These rules pertain to semantics, syntax, and phonology.
Chomsky (1957) was among the first to describe the creative aspect of grammar: A finite set of rules is used to create sentences appropriate to an unlimited number of situations. The user is "capable, in principle, of generating any of the sentences from that set and at least some of them are likely to be new or novel but at the same time appropriate to the situation, whatever that situation may be"... (Palermo, 1978, p. 31). This characteristic is true for all human languages and thus is a universal one. Genera-

tive linguistics describe this universal characteristic in detail.

Being creative is one of the most striking differences between human and animal language. D. Premack (1970, 1971, 1975) is a researcher who tried for many years to teach language to chimpanzees. While I was in the United States, he was a guest speaker at the Center in the university where I studied. In discussing his work, he told us what it was like when his grandchildren would come to visit him. Each time, they would ask him to play the following game:

He was to tell them a familiar story. At some point, he would change something, e.g., the sequence of actions, the name of a person, the kind of animal, etc. When he did this, there would be a great outburst, "No, Grandpa, not like that. Like this . . . ." Then they would be quiet again and listen with great attention, waiting for the next occasion when he would mix up an event.

"You know," Premack said, as he finished his speech, "playing such games is only a characteristic of human beings. It is not one of the chimpanzee".

Sentences consist of a surface structure and a deep structure (Chomsky, 1957). Transformational rules relate the deep structure of a sentence to the surface of a sentence.

Simple sentences involving a minimum of transformations can be called kernel sentences. One can assume that these sentences are closest to the deep structure. Fillmore (1968), representing generative semanticists, pointed out that syntactic-semantic relations always include an action with an actor or agent. Often, there is also an object included, like in the sentence: The dog bites the girl. Events expressed by the actor and the action are causal and lead to changes.

Transformations and embedment change the kernel sentence to one that is more complex. The actions can lead to different questions: Who did it? To whom did it happen? Where did it happen? What changed? Fillmore (1968) introduced the term "case" to express this kind of syntactic-semantic relations determined by the action. This became the foundation for "case grammar".

Generative grammarians assume that deep structure is universal and innate. (For more information, see Palermo, 1978.) Others (Piaget, 1963; Brown, 1973; Edwards, 1974; Bloom & Lahey, 1978) argue that deep structure is rooted in sensorimotor experience.

Considering the importance of events in the center of syntactic-semantic relations, which include actor, action and cause-effect, it reminds us of our description of problem solving events in daily living. These are the root of development for different performances which lead to the assumption that problem solving events in daily living are also the root for the development of deep structure with universal character. Children all over the world continually try to solve problems in their daily lives, applying a finite set of rules to an unlimited number of situations – the creative aspect of deep structure!

Transformation rules relate deep structure to surface structure. Therefore, transformation rules can also be considered as being related to problem solving events in daily living. Who did it? Where? How? What? As a result, we can assume that *problem solving events in daily living are also the root of transformation rules*, and therefore, the root of semantic-syntactic relations.

This model of *problem solving events in daily living as the root for language* is also illustrated by the different observations which were reported involving normal children who are learning language.

The question arises: What happens to those children who fail to solve problems in daily living because of their perceptual disorder? When the model is realistic, we can expect these children to fail in surface structure as well as in deep structure.

2.3 When the Root is Sick...

Given that semiotic performance is strongly related to problem solving events in daily living (see Part III A, paragraph 3.2 "The Model of Development"), it is expected that children with difficulties in the production of problem solving events need a *longer time* to reach the semiotic level. The length of delay depends on the kind and degree of perceptual disorder. The delay will be observable in different semiotic performances, such as, symbolic behavior and language. Besides the delay, we can also expect a *deviancy*.

2.3.1 ...the Symbolic Behavior Is Deviant;

Perceptually disordered children are delayed in using *forms* which are *available* to express content that is not inherent in an actual situation. When they do begin, specific *deviancies* can be observed when the performances are analyzed carefully.

D., 5 years and perceptually disordered, comes to the Center for an evaluation. His mother is asked if he plays symbolically.
She answers, "Yes. He plays with puppets. For instance, he has an alligator try to devour Kasperli".
"How does he do that"? I ask the mother.
The mother responds, "He takes Kasperli and the alligator, and holding one next to the other, he says, "Alligator devours Kasperli".
"And then"? I ask. "What happens right afterward? Does the alligator devour Kasperli? Aaugh! Like that, with the mouth wide open, so that Kasperli shouts and cries and disappears into the alligator's mouth"?
The mother hesitates, thinking the question over silently to herself, and finally says, "No, nothing happens".

It seems, therefore, that D. can *produce semiotic performances*. That is, he uses forms to represent a content, but the content consists of something he has seen in the past, such as, the alligator next to Kasperli. He also knows what to say in this situation: "Alligator devours Kasperli". Thus, D. can use what he sees and adds what he hears. The real action of devouring and the tension which should go with it are missing in his representation. We interpret: D.'s representation of *content is poor*. The movement of putting the alligator next to Kasperli is a simple performance. The content is mainly visual and auditory. Tactile-kinesthetic information is hardly included.

Such poor content is also expressed in the next example:

R., 11 years, is perceptually disordered. His mother tells us how he always plays the same game. He has a toy airplane which he moves from one place to another in his room. While doing this, he calls out the names of cities, "Athens...Rome...Zurich"! He had taken an airplane trip with his parents during the summer, and now he represents that trip. At first, some of his friends participated in the game, but they no longer come to play with him. The game has become too boring for them. They say they cannot understand his game.

The forms that perceptually disordered children use for their symbolic playing are already *in the environment*. This makes them similar to normal children, but if we bear in mind the other possibilities for using forms, the differences are quite noticeable. *Seldom* can we observe them using their *own bodies* to represent activities. The *construction of forms* to represent problem solving events is *missing*. Perceptually disordered children frequently learn to construct, but they usually do not utilize their constructions for representing events.

K., 5 years and perceptually disordered, puts tubes together in his spontaneous play. He fastens them to faucets and watches how the water flows through the tubes.
When we first observed the action, we were

impressed by his skillful manipulations, but watching his activity over a longer period of time, we became conscious of the monotony of the repetitions (see Part II, paragraph 2.2.2 "And Where Is the Neighborhood"?).

These kinds of constructions usually include short cause-effect activities, such as, putting the tubes together. Often, perceptually disordered children are oriented toward a visual impression which they try to reconstruct.

N., 12 years and perceptually disordered, gathers sticks, bars, and Play Dough. He rolls the Play Dough into stick-like shapes. He takes those shapes and quickly makes an alligator. As soon as the shape is made, he destroys it and starts again.

L., 6 years and perceptually disordered, draws lamps on every sheet of paper she can find. She doesn't care if a drawing is already on the paper. We get the impression that her lamps are without content.

The behavior found in normal children who use *symbols* to explore a *new* and *difficult situation* is missing in perceptually disordered children. When situations change and become difficult to be mastered, children who are perceptually disordered panic, retreat, or begin to ask one question after the other (see Part II, paragraph 3.3.3 "What Happens When I Am in Search of Tactile-Kinesthetic Information"....).

In *picture recognition* perceptually disordered children, like normal children, recognize objects before they recognize actions represented in the pictures. What is strange, though, is that they stay in this period for a long time. They recognize represented events only after a long delay, and even when they begin to do so, the interpretation of the *events remains fragmentary* and deviant because they rely on a high degree of visual information.

At our Center, there is a picture book which includes an illustration of Indian life along a river. The river winds through the landscape, from the foreground to the background. There are many stones in the bed of the river. By chance, one of the stones in the background is painted so that it appears to be right above the head of an Indian in the foreground. For normal children this does not present a problem, for they see the stone as part of all the other stones in the river. This is not so for perceptually disordered children. Many of them will point out that the Indian has a stone on his head.

A nonverbal test we frequently use includes a problem which consists of a drawing with missing parts. For example, the children must choose from a group of small pictures and indicate which ones fill in the missing parts in the drawing. Some of the pictures fill the missing parts in a "visual" manner, and others fill them in by referring to the actions represented in the drawings. It is interesting to note that perceptually disordered children often choose the visual pictures and not those which refer to the actions.

Usually, pictures can be *named* correctly, but the children cannot act out what the picture represents when asked to do so.

K., 11 years and perceptually disordered, names a picture correctly: "The dog is biting".
I ask him to pretend he is the dog. He stands up but doesn't move, appearing to be helpless. It is obvious that he doesn't know what to do.

This reminds me of the example of Kasperli and the alligator. D. could name the situation: "Alligator devours Kasperli," but the devouring action was missing.

What is visible in a picture is named, e. g., the position. There is no reference, though, to the action represented by the picture (which is not visible). This can be taken as a sign of an exclusive dependency on the visual information and of a lack of tactile-kinesthetic information.

Most perceptually disordered children begin to *draw* but it is greatly delayed. Once they do begin, they exhibit difficulties. Often their drawings are not like those of normal children. They draw constructions, like buildings and train systems, but actions are seldom represented.

The first level of "scrawling without representing" is often not observable in perceptually disordered children. Once they start to draw, they may produce a confusion of forms. They may name the drawing afterward but often change the content while naming it.

Some perceptually disordered children are oriented so much toward visual information that they begin with "visual realism". That is, their first drawings are on a level that comes last for normal children (Piaget & Inhelder, 1956). Some are so extremely visual in their representation that they begin to draw in perspective around 5 years of age.

2.3.3 ...the Deep Structure and Surface Structure Are Deviant

When we ask parents about the babbling of their perceptually disordered children, they often hesitate and finally report that their child was extremely quiet – except when crying. Babbling? No. They do not remember it. Despite the missing stage of babbling, perceptually disordered children will begin to speak after a delay.

Perceptually disordered children, like normal children, acquire language in a *creative* way. Both extract forms from the surface structure offered by the speaking environment and affix a content to them that reflects their experiences. Thus, perceptually disordered children do not learn language by imitation either. The process is far more complex.

For a long period of time, the *understanding* of verbal communication is restricted to actual situations. In this case we must deal with signal comprehension and not language comprehension (see Part III C, paragraph 2.1.4 "The Picture Is a Symbol").

When semiotic comprehension appears, it is fragmentary for a long time. The mother of D. complains that he listens, and even watches her when she tells him something, but she is never certain if he really understands what she is saying. In another situation, a child wants to know where they are going. Someone explains, and the child is quiet for a few moments but then repeats the same question. It is difficult to tell stories to perceptually disordered children. Even as late as kindergarten age, they may still have difficulty sitting quietly and listening when a story is being told.

Production of language can be delayed for several years. This can cause deep concern for the family and others in the child's environment. "When will my child speak"? they ask.

Often, once these children do begin to *speak*, the rate of speaking increases rapidly and becomes excessive. It is as if perceptually disordered children compensate for a lack of other activities by speaking (see Part II, paragraph 1.2 "They Talk Incessantly"). Poor articulation is another concern. Many perceptually disordered children receive speech therapy. Even so, at an older age their sound production in a stressful situation deteriorates faster than it does for normal children.

Of concern is the poor *syntactic* structure. Reversals are observable, e. g., "I going was", regular forms of past tense verbs are generalized, and irregular forms of verbs are missing. Sentences are often long and appear to be correct but have the characteristics of phrases. A closer analysis reveals that subject-action references are inaccurate. Kernel sentences are seldom, if ever, produced. The frequency of phrase production and the lack of kernel sentences appear together with semantic-syntactic difficulties. Transformational rules are often used incorrectly. The structure – subject-action-object – is missing. These observations indicate the difficulties, not only on the level of surface structure, but also on the level of deep structure. This interpretation is strengthened by the fact that it takes perceptually disordered children a much longer time than normal children to report what happened at school.

Frequently, we observe *semantic confusion*.

S., 6 years and perceptually disordered, often asks for sugar instead of flour when he is baking. He also confuses the names of tools when identifying them, such as, the hammer for the saw or drill.

Likewise, we can observe their difficulty with finding a verbal expression – a problem of *verbal retrieval*.

These observations depict the existence of difficulties with surface structure as well as deep structure. They support the assumption that the problem stems from a common failure – the root is sick. The question arises: When the root is sick, what can be done?

2.4 What Can Be Done?

Our research data allow us to assume that the root of development which has been described (see Part III A, paragraph 3 "How Can We Represent Development"?) is also the root for developing semantic-syntactic relations. This root consists of experiences in problem solving events in daily living. Day after day normal children are active with such events. The same is true for perceptually disordered children. Even as an adult, young or old, normal or brain damaged, we have to solve problems that arise in daily life as long as we are alive. This means that there is *no age limit for therapeutic intervention*.

2.4.1 Something Can Always Be Done

Two examples illustrate this:

L., 18 years, suffered a severe head trauma.
After she was in the hospital for 6 months, the medical director called her parents and told them that they should look for a nursing home for L. There was nothing more that the hospital could do for her.
Her parents were filled with consternation, but after reflection, they decided, "We will take L. home and have her become a part of our daily lives. We will not give up on her"!
The parents followed through, looked for help, and sought counselling. They sustained L. and helped her fight to live her life.

It is five years later. L. is still handicapped, but she can walk and speak. A year ago, she even travelled to Paris with her parents where she directed the trip and did the translating. Though there is work to be done – she still needs to become more independent in walking and in solving the problems of everyday life – she is leading a meaningful life.

This progress was possible through the use of problem solving events! Physical and occupational therapy, water therapy, the hard work of her parents, and the care and concern of others in her environment were used to repeatedly intervene, with her and for her, in *daily living events*. Problems of daily living were solved with L. in the past, and will still be solved in the future – over and over again! Thus, something can be done – at any time. This is an *important challenge for those of us in the environment*.

M., 8 years and perceptually disordered, is still not at the semiotic level, and consequently, she has no language. Her parents come to see us because they are desperate. They have been told that a child who does not speak by the age of 6 years will never learn to speak.

I was dismayed. How could anyone maintain such a belief? There are no research findings to confirm such a theory, and our studies contradict it. We have children in our longitudinal study (SNF nr. 3.929-0.78), who have had therapy for several years and who began to develop language when they were between 12 and 14 years of age. Gardner (1983) described a girl who started to speak during her second decade of life. To emphasize again: *We do not yet know about any age limits for developing language.* The main condition for progress is: tactile-kinesthetic experiences in problem solving events, today, tomorrow, and possibly, for many years!

2.4.2 First Comes Tactile-Kinesthetic Input from Problem Solving Events of Daily Living and Then Comes Representation

In this Part III of the book, we reflected at length on how to guide someone through a problem solving event. Now we need to answer the question: How do we represent such events?

Before beginning with representation, we must make certain that the *perceptually disordered child is functioning on a semiotic level.* To do this, it is important to evaluate semiotic comprehension in addition to semiotic production.

In the first part of the chapter, we described the behavior of children who are on a semiotic level.

- We can *communicate* with them about a *future* event.
- We can talk to them about an event which is *past*.

In both instances, the child listens with understanding and does not expect the event to happen at that moment.

T., 14 years and perceptually disordered, is in a small group at a school for perceptually disordered. The children have just returned from swimming. The teacher begins to talk about the swimming experience as a past event and also mentions that they will return to the pool the next morning. Immediately, T. hurries upstairs to fetch his swimming suit and then stands in the hallway, waiting to leave for the pool, as if they were going to do go swimming again right then.

T. is still not functioning on a semiotic level. He understands verbal messages within an actual situation but not if they refer to the future or the past. This means he has signal comprehension but does not understand signs. He expects the event to happen now.

Other indications of semiotic behavior are:

- *If we* represent *an event that has happened by a drawing and the child watches* attentively *and understands.*
- *If the child is conscious of the fiction* or the "doing-as-if" of symbolic play. For instance, we can be looking at a picture with a child where the picture represents a child who takes a ball away from another one. If we behave as if we are angry, and scold the child in the picture, does the perceptually disordered child at our side realize that we are only pretending to be angry, or does he or she have the impression that we are really angry and react accordingly?

As soon as we are certain that the perceptually disordered child or adult understands semiotic behavior, we can begin to represent problem solving events which he or she has performed or has been guided through.

If we work with a group of perceptually disordered children or adults, we may have to ask them to perform a semiotic production (see the next example of N., 9 years), but when we can work on an *individual* basis, we need to consider several *conditions.* These conditions "hold true" for perceptually disordered children and adults, including those with aphasia. (Aphasia is a loss of language.)

The most important condition is that we work with the perceptually disordered children and adults on *their level of comprehension and not on their level of production*. For some therapists and teachers, this is a difficult condition to fulfill. It contradicts their usual procedure and means that they must focus on spoken or graphic forms a perceptually disordered child or an adult cannot perform. An example is: I sit down with a perceptually disordered person to represent the guided event of problem solving we just finished. I refer to the event by drawing or by speaking or using gestures or even written words or sentences. It is assumed that he or she could not produce such drawings or spoken forms or gestures, nor write words or sentences using a similar complexity. In other words, I produce forms which are above the level of production but within the level of understanding (see Part III B, paragraph 3 "I Understand Problem Solving Events of Daily Living").

Before a problem solving event, I do not focus on representation at all, or I make it very brief. This is because at this time perceptually disordered children and adults have not yet experienced the content of the event and are still missing the corresponding tactile-kinesthetic information.

I represent the event *after* it happens. Two examples demonstrate this procedure. (We emphasize again, for the respective children or adults, the selection of the kind and complexity of the forms should correspond to their level of comprehension and not to their level of producing language.)

In school, N., 9 years and perceptually disordered, often draws when she is left on her own.

The drawings are produced without reference to tactile-kinesthetically perceived problem solving events. They all look alike. Drawing 1 is an example (see p. 287). The drawing consists of a profusion of forms which fill the page without connection. When we ask N. about the content, she points to individual forms and names them. While doing this, she often points more than once to the same form and indicates a different content each time.

Most confusing are those forms which represent a human being. On drawing 1 she has represented herself in the upper left corner.

When N. is guided through a problem solving event and afterward is asked to draw, her drawings are different. Drawings 2 and 3 are examples (see pp. 288-289).

Drawing 2 represents the situation. In the foreground, she drew herself with some letters on the front of her sweater. What a difference this drawing is as compared to the self-representation in drawing 1. In the background she sketched her teacher who had guided her through the event.

Drawing 3 represents the actions of the problem solving event. She prepared fennel for cooking and was guided while performing this task. She tried to break the stem of the fennel by hand. This required great effort.

Notice that she drew the form of the fennel in relation to the tactile-kinesthetic information she probably received when working on the fennel, e. g., the roundness of the form and as a contrast the stem sticking out of the round vegetable. Of further interest is the representation of the people involved: She places herself in the foreground – this time with arms and hands shown; she correctly places the letters on the pocket of her sweater; and in the background, she shows her teacher.

Mrs. O., 40 years, who is brain damaged and has aphasia, received therapy in a clinic for a long period of time, but she is now at home. She can once again take care of her household, and she continues to travel by bus for therapy at the clinic. However, now it takes her a much longer time and more energy to perform daily activities than it did in the past. She can report simple events, but difficulties appear when she has to search for a specific expression. She often cannot retrieve the expression, becomes tense, and her verbal performance breaks down. She tries to write with her paralyzed right arm, but can hardly do it correctly, let alone read it once she has tried.

Drawing 1

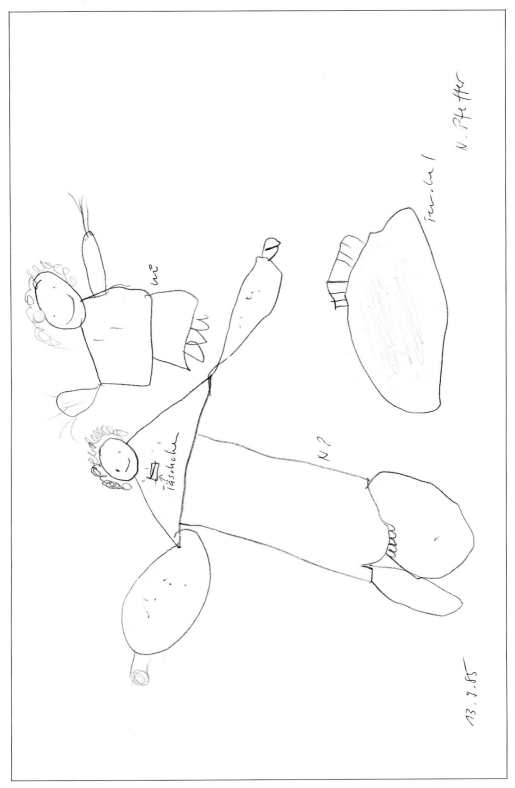

Drawing 3

I am guiding Mrs. O. as we begin to divide a banana. She is very tense. Her right side is especially stiff and can barely be moved. After about five minutes, I interrupt the task for a break - not because Mrs. O. is tired but because I am. We were only able to reach out for the banana across the support of the table, but no to take it off.

I take a sheet of paper and write: We will divide the banana. We are reaching out for the ——— . Mrs. O. tries to write the word "banana". She pronounces it spontaneously; she even finds the word in the first sentence. She tries to write it, but the letters are hardly legible and the word is spelled incorrectly. I guide Mrs. O. and we write the word. Now we continue the guided event of dividing a banana. The guiding becomes easier. Mrs. O.'s body tone decreases. After 10 minutes, we interrupt our effort. Mrs. O. smiles and says, "I feel so good". She points to her paralyzed right side, shoulder, and arm.

An hour passes in this manner. We perform a part of the guided event. I become tired. During the break, there is a short verbal representation of some aspects of what we had done. The written and oral forms used were on Mrs. O.'s level of comprehension but above her production level, and their contents were connected with the tactile-kinesthetic experience of the event. It is during these representations that Mrs. O. spontaneously performs some small parts of the linguistic task. After we have tried to break a piece of the banana by hand, I write another unfinished sentence, and we read it together, "The banana is ———".

"Mushy," Mrs. O. answers. I might not have found a better expression to describe what we had felt when breaking off a piece of the banana. The banana was getting soft, slippery, sticky, wet, and yes - mushy.

At the end of the hour, we had broken off three pieces of the banana (only that much). On the sheet of paper there is a small text. In the last sentence I leave out the word banana and let Mrs. O. write it. She does. I watch her write. I cannot believe it. The word is written neatly and legibly. I can hardly differentiate between what she has written and what I have. The word is correct and Mrs. O. is amazed as she sits there looking at the result of her efforts. She can barely understand it.

I realize that this kind of a performance will not be possible for her to repeat again in her own home, but slowly, very slowly, and with numerous tactile-kinesthetic experiences in problem solving events during therapy, her body tone will become more normal. She will relax more easily and more frequently; her writing will improve; and she will retrieve verbal expressions more quickly.

These examples are only two of many that demonstrate the importance of tactile-kinesthetic experiences in problem solving events in daily living. This principle is not only valid for adults to rebuild their performances, but also for children. Thanks to such tactile-kinesthetic experiences, progress will be noticeable in the surface and deep structures of language and in the comprehension of visual and auditory forms. As time progresses, increased interiorization of these experiences will also improve production performance.

Therefore, it cannot be stressed enough: *Perceptually disordered children and adults (including aphasics) primarily need tactile-kinesthetic experiences with daily events in solving problems. Only after providing them with such events will we use forms to represent the tactile-kinesthetic event.* It is here that we can apply what we have learned as teachers or speech therapists. It is here that we can become involved in building up language, practicing articulation, reading, and developing writing and arithmetic skills.

One last remark: Language is an important means of *communication*. We can apply this functional aspect of language when we represent an event with children and adults. For instance, we represent an event graphically for the purpose of communicating it to someone else, e. g., parents, siblings or friends. We may also keep a record of the event so that we can refer to it in the future by reading it or repeating it.

To Summarize:

This chapter included some difficult reflections. The problem of representation leads to complex questions.

Semiotic behavior is only a characteristic of human beings. We briefly described this human performance and contrasted it with the attempts to teach language to a chimpanzee.

We observed *normal children* who were beginning to discover symbols and language. They begin by using available forms and afterward, use forms which they construct themselves. We described how they take forms out of the environment connected with problem solving events, and how they use them to represent problem solving events they have experienced in the past. In this way, children begin with semiotic performances, e. g., symbolic play behavior, picture recognition, drawing, comprehension of spoken language, and speaking.

We reflected again on the root of development - tactile-kinesthetic interaction with the environment in the form of problem solving events. This was to emphasize how the root relates to the acquisition of deep structure and surface structure.

Observations of semiotic performances in *children and adults who are perceptually disordered* emphasized the same point of view. They are deviant in language as well as in other semiotic performances, not only on the level of surface structure, but also on the level of deep structure.

These deviations, like the others described, can be related to the disordered root, i. e., the disordered experiences of interaction with problem solving events.

With these remarks, the title of this Chapter 2 "Return to the Problem Solving Events of Daily Living - Then Comes Representation," should have become more comprehensible. First comes the acquisition of deep structure. Only after that working with surface structure will become meaningful, i. e., working with forms. This is true for the many kinds of forms which can be used for representation: spoken forms; pictures - drawing; reading - writing; arithmetic, and symbolic play.

In the introduction to the book, we limited our reflections to the "discovery" of language. How language will further develop is a problem that is beyond the scope of this book.

For the further development of language, however, we remind you: *Work on the root and not on the branches!*

D. Conclusions

Though the conclusions are far reaching, we can provide only short descriptions within the framework of this book.

1 Problem Solving Events Can Be Considered the Root of Development

This statement, with all of its ramifications, applies to all of us.
Working on the root means stimulating development, reconstructing disordered performances, and acquiring new performances. This method of working is applicable when we have to deal with development, learning, and rehabilitation.

1.1 Possibilities of Application Have a Variety of Forms

With children: Those with normal development as well as those with any kind of abnormal development are included. There are many aspects to reflect upon. What does the environment offer for experiencing the tactile-kinesthetic interaction necessary for learning and development? The immediate environment includes the home and the surroundings, such as, gardens, streets, buildings, fields, woods, meadows, etc. What kind of activities are possible in the immediate environment? Do the children have the opportunity to explore them tactually, to act upon the surroundings, and to create changes in their environment? Can they perform causative actions and experience their effects? What kind of problems in daily living situations can they experience within the family? Within the school? How often is tactile-kinesthetic interaction replaced by audio-visual pseudo-interaction as in the case of the television? Where is the possibility to feel the interaction? To meet resistance? To experience a regularity of changes in resistance?

With adults: Primarily the brain damaged adults are included. But it is also important to consider working with the geriatric population. We should be working on the root with both groups – provide them with intensive tactile-kinesthetic experiences when solving problems in daily living. How many performances could be stimulated? How many experiences of the past could be retrieved and made usable again? How much would one need to change in their immediate environment in order to bring more of their past daily life situations back again? How much change would be needed in homes for senior citizens? In nursing institutions?
In reflecting on adults, we must also include normal people – ourselves. We are usually tempted to allow habits to exist and expand so much that we become rigid and rarely think about causes and effects. Are we conscious of the Wirklichkeit as it is before we change it? Do we consider the effects we create by the actions we perform? Do we still feel the changes we are eliciting in the Wirklichkeit? Or do we live in an audio-vi-

sual world, and avoid encountering changes in resistance? Are we concerned about our root? Are we careful that they remain strong and, as a result, continue to grow?

The statement about working on the root leads to another important conclusion: Transmitting tactile-kinesthetic information through guidance is also possible with *those who are severely disordered.* Recall the discussion in Part II, "Failing in the Wirklichkeit," paragraph 1 "Those Around Them Notice: They Are Deviant". We described perceptually disordered children and adults as being very difficult to have around us because their behavior appears aggressive. As a result, they are quite often referred to psychiatric clinics or nursing homes. Other brain damaged patients withdraw, and their behavior is judged as having changed or deteriorated. There are those who are so spastic that they cannot move spontaneously, and there are those who are judged by others as being without contact, i.e., autistic. The list could go on and on. Unfortunately, there has been no systematic investigation of these severely handicapped patients from the point of view of working on the root. We only know about some individual cases where those around them – their parents, another member of the family, or therapists – were involved in a special manner and applied our method. These individual cases support our formula: It is possible to work on the root with those who are severely disordered. They indicate further that there must be many more cases like these where an individual might be helped with this kind of therapeutic intervention.

Problem solving events are related to *daily living.* This leads to other special requirements for therapeutic work.

1.2 The Application of the Therapeutic Model Involves a Whole Circle of Persons

The best place for daily living situations is *with the family.* Whenever perceptually disordered children or adults live with their families, it is important to get *cooperation* from the individual members.

This can present a dilemma, especially in the beginning. The difficulty may be created by the stress situation within that family. A therapist does not often realize the severity of that stress. They are also not conscious that a healthy person under stress shows reactions which are similar to the ones observed in those who are perceptually disordered (see Part II, paragraph 1.2. "They Talk Incessantly"). Unfortunately, this phenomenon cannot be discussed in detail in this book. We can only suggest that, as a consequence of stress, there is an overflow of tactile-kinesthetic information which can overload our capacity (see Part II, paragraph 3.3 "The Limitation of Capacity".) Thus, we may talk excessively, complain or demand actions from other people, and become very tense, too.

When this kind of behavior is observed in the families of our perceptually disordered patients, what can we do?

Our experience has revealed that the less the family members know about how to deal with the perceptually disordered person the more pronounced their stressful behavior is. Therefore, it is important that the family becomes able to work with the patient as soon as possible. This requires that family members become more conscious of the kind of tactile-kinesthetic information that can be transmitted in a problem solving event. To achieve this awareness, we guide the member of the family, and they guide us. Then we apply the same kind of guiding to the perceptually disordered child and adult in that home. When this is accomplished successfully, the family's stressful reactions usually decrease.

Counselling and working with the members of the family requires much of the therapist's time. The same requirement should be met in schools, i.e., working with the families should be part of the curriculum.

In order to intensify work with the family, "family weeks" have been found to be a very rewarding possibility. About four families – including the parents, normal siblings, and the perceptually disordered child – spend one week together. Each family lives in a tourist apartment in order to allow for family life. Under professional guidance, the special problems of the perceptually disordered child, the normal siblings, and the parents – including problems of their daily living – are discussed.

It is essential that the *members of the family*, especially the mother or the person responsible for the perceptually disordered child or adult, are partly *relieved* of other family duties so that time is available to deal with the additional responsibility. In Part III B, paragraph 3.3.3 "Feeling Is Very Difficult," we emphasized the necessity of calmness and sufficient time in order to focus complete attention on the guided person. The time spent in guiding does not have to be long – only a few minutes. Even this may be impossible, though, if a mother has to take care of other children at the same time.

Often perceptually disordered persons are in an *institution* or a *clinic*. Unfortunately, as we pointed out in Part III B, paragraph 3.3 "Guided Through Daily Living," many institutions and clinics offer few daily living experiences. If these circumstances exist, one should inquire about the possibility of changing the conditions or changing the institution or clinic. In any case, it is necessary to get the *cooperation and coordination of the different persons responsible for working with the child or adult.*

Coordination of all therapists – occupational, physical, recreational, and speech – is necessary. For example, while the occupational therapist "guides" the patient through a problem solving event in daily living, the physical therapist helps to get the patient into specific positions for moving, such as, standing and then walking during the event – performances involving exercises which the physical therapist would probably be worked on separately. The nurse on the ward or the educator of the group could do some guided events within his or her group activities. The speech therapist refers to the guided event by representing it with the kinds of verbal forms that he or she would be practicing anyway. These same events could also be used by the school teachers for their work.

2 Our Knowledge About the Root of Development Is Still Limited

Our research projects begun a *few years ago* with the financial support of the Swiss National Science Foundation. The subjects of the longitudinal and cross-sectional projects included children with perceptual disorders, normal development, hearing impairment, and brain damaged adults.

The research covers many years. We have compared it to a hike up an unknown mountain. We said that we have not yet arrived at the top, but to reach it, we must continue to search for the right path. Many questions have arisen during the hike – some of them could be answered; others could not.

Therefore, the research has to be continued and extended.

2.1 Extension of Longitudinal Research Is Needed

The longitudinal projects should be *extended to other groups* of children, such as, those who are cerebral palsied, visually impaired, or mentally retarded. In addition, adults,

e. g., brain damaged, geriatric, and psychiatric patients, should be included.

These studies should be further extended as to the *kind and number of observations*. New data in our research encourage additional observations and an extension of existing observations. Also, collected data of the *longitudinal studies* need to be *analyzed, described, and published*.

2.2 Extension of Cross-Sectional Research Is Needed

Experimental studies of the *perceptual performances* of normal, perceptually disordered, and visually and hearing impaired children should be expanded to other groups of children and adults (see preceding paragraph 2.1).

In addition, *other kinds of performances*, which have been discovered to be critical, should be examined cross-sectionally, e. g., nonverbal problem solving activities.

Also important is the *analysis and publication* of data already gathered in the cross-sectional studies.

The study of *other problems* should be initiated to fill the gaps of knowledge pointed out in this book:

- Investigation of *tactile-kinesthetic perceptual* processes and their development.
- Investigation of the development of *bimanual performances* and spatial concepts with special reference to the *support*. For example: How do normal children develop from using one hand to using two hands? How do children acquire the "harmony of moving 10 fingers"? How do children switch from embracing to taking off and putting back? How do children acquire notions of direct neighborhood?
- Studies should be undertaken to describe the development of nonverbal problem solving processes in normal children and their relationships to the deep and the surface structures of language.

Each of the open-ended questions applies to the children of the other groups mentioned (see preceding paragraph 2.1), children with different disorders.

Besides such studies of development, the question about the *regression of already developed performances* arises. How can we evaluate the regression of some of the performances in adults described in the research?

Many questions are still unanswered. Much more research is needed if we want to continue on our hike. We have only been able to briefly describe this need.

It should be evident to anyone that this kind of research is only possible with outside help – supportive, but especially financial. In addition, the urgency of the questions raised requires the interest and the cooperation of other places responsible for research.

Closing Remarks

While reading through this book, some of you may have done it carefully, savoring each line, while others may have just skimmed some of the chapters. Many examples may have given you pleasure. Others may have gotten "under your skin". Some examples may have been contradictory. It is my hope that others may have caused you to reflect or to stimulate your curiosity.

We began the book by "being on a mountain" and considered situations: Here I feel the world near me – there I see and hear the world far away.

We pondered several questions: They were found in numerous examples on many pages of the text. They were asked over and over again, often formulated differently.

We asked: **Why do the surroundings limit our moving around?**

We observed *that normal children* move, are moved, and touch. They find resistance which allows them to recognize the existence of the world.

We asked: **Why do we search for a broader space, and at the same time, for a restricted one? For social life, and at the same time, for solitude? For freedom, and at the same time, for resistance? Why are there such contrasts?**

Again, we observed that *normal children* move and are moved; they touch and find resistance; they embrace in order to hold and to take. In this way they change the resistances between their own bodies, the support, and the side. Children feel the regularity of these changes and begin to form the rules of touching. A limited set of rules allows them to grasp the changes in an unlimited number of situations. To do this, children – and adults – need contrasts. As an adult, I can estimate the broadness of space only when I also experience its narrowness; know social life only if I also know about solitude; feel freedom only if I also feel limitations. Contrasts and changes are, therefore, the preconditions of perception and of knowledge.

We observed *children and adults* who are *perceptually disordered.* They, too, move or are moved, touch and find resistances. They, too, can discover regularities and form rules of touching. They, too, need contrasts and changes. However, because of their perceptual disorder, they continue to perceive only the strong, maximum changes in resistance far beyond the ages when normal children can perceive minimal changes in resistance without difficulty. Consequently, perceptually disordered children and adults try to elicit maximum changes in resistance over and over again. Therefore, their tone increases rapidly and they push with full force. Those around them judge them to be aggressive, self-destructive, tactile-defensive, etc. The perceptual disorders are not only the cause of the search for maximum changes in resistance, but also the cause of a serious lack in tactile-kinesthetic information. As a result, changes of situations are not perceived at all, or are only perceived fragmentarily. This prevents those who are perceptually disordered from interacting adequately with the environment and receiving the necessary tactile-kinesthetic experiences of interaction. *The root of development is disordered.*

We asked: **How do we get to know about the world that is near? How about the world that is far away?**

Here *normal children* were described again. They move and are moved, they elicit movements - causes? And their effects? They touch, embrace, take off and release. The changes in resistance they elicit and feel are of a great variety and include their own bodies, the support, and the objects and people being on that support. The regularity of such changes which are perceived tactile-kinesthetically are the basis for the rules of acting upon. Children begin to bring order to their tactile-kinesthetic experiences with causes and effects by using these rules, such as the rules of taking off and of neighboring relationships. They apply them to innumerable situations of daily living and soon begin to learn about the Wirklichkeit as it is.

At first, the world at close range is perceived tactile-kinesthetically. Then the world farther away becomes included by adding visual and auditory information to tactile-kinesthetic information. At the same time that children acquire tactile-kinesthetic experiences about the Wirklichkeit, they acquire knowledge about their own bodies - contrasting aspects of the same events. What they feel with two hands and different parts of the body becomes a unity, and the moving of the 10 fingers, a harmony.

With the increase in knowledge about the Wirklichkeit the child begins to be able to change it, and with that, to solve problems in daily living.

We discussed those *children and adults* who are *perceptually disordered*. The world that is near does not become familiar even though They, too, recognize that they can elicit movements in their surroundings by touching them. However, they continue to need the maximum changes in resistance, and it takes them a long time to be able to take an object off. It takes them so long that, in the meantime, they may learn to walk. They, consequently, do not include the notion of the support in their experience with taking-off and with cause-effect relationships in their surroundings. The deviant performances which result from these kind of abnormal experiences were discussed. These children focus on information from a single hand, the harmony of moving 10 fingers is missing, and relationships of neighborhood become deviant. They do not get to know about the Wirklichkeit!

We asked: **How do we get from perception to language? What does the Wirklichkeit have to do with that problem?**

Normal children were described. They begin to solve problems of daily living through increasingly more complex situations and through changes in their surroundings. Tactile-kinesthetic experiences of such events become interiorized - deep structure is formed. Such interiorization reaches a critical degree so that language is discovered, i.e., children *discover* the forms which are presented by the environment, the surface structure, and at the same time, the possibility of using these forms to represent the interiorized tactile-kinesthetic experience with problem solving events of daily living. *Language begins to develop* in a direct relation to the growing of the root - the experience with tactile-kinesthetically perceived problem solving events.

The *failures of perceptually disordered children and adults* were described. The lack of adequate tactile-kinesthetic experience with problem solving events interferes with the normal acquisition of deep structure and surface structure and perceptually disordered children are late in beginning to use the forms presented by the environment. They try to communicate with them, but the usage and development of surface structure, as well as, the semantic content - expression of deep structure – appears deviant. All of this is an *indication of a disordered root.*

What c*an we do?*

We pointed out that we have to work on *the root* and not on the branches. We emphasized that we cannot help perceptually disordered children and adults simply by waiting until it happens or by enriching the environment (Riesen, 1975). We noted that perceptually disordered children and adults also form rules; they learn to recognize cause and effect relationships. They are competent in those rules, but their lack of information prevents their competence from becoming adequate performance. The most they can learn are habits because the lack of information inhibits them in adapting their productions to changing situations.

We concluded: **Our duty is to provide perceptually disordered children and adults with better information when solving problems in daily living events.**

We do this through guidance. By guiding we provide tactile-kinesthetic information. Tactile-kinesthetic information about changes in resistance within problem solving events is perceived whether such persons are guided or whether they perform the movements themselves. Tactile-kinesthetic information about changes in resistance which are inherent in the problem solving events allows perceptually disordered children and adults to experience interaction with the Wirklichkeit. This allows for changes in behavior – for learning.

In this way we work on the root with perceptually disordered children and adults. We can expect that with increase in interaction experiences in changing situations the understanding of problem solving events is extended. With increase in interiorization development will progress, productions will improve – some slowly, some more quickly; some more, some less. All of this will depend on the degree of the disorder, the possibility of help, and other factors.

In this way, we hope that the perceptually disordered children and adults for whom we are responsible can find the way to *their* mountain, can continue on the path, and do not have to remain stagnant or regress.

Glossary: Wirklichkeit

I have used the German word "Wirklichkeit" as well as its concept throughout this book. Its meaning and use are crucial to your understanding my philosophy of the development of children.

The dictionary translates Wirklichkeit as "reality". But reality doesn't capture the entire concept. Reality comes from the Latin word *res, a thing*. The connotation of "reality" traditionally points to a way of thinking about concept formation where an object *per se* is focused on as "something" one can see, hold, and manipulate in an absolute way.

However, our observations of children underline that their activities are not centered on objects *per se*, and children appear not to be interested in an object *per se*; we can assume that their development is not simply "object oriented".

The use of "reality" does not take into account such kinds of activities, and thus would be misleading in the confines of this book. "Wirklichkeit" as a German concept describes this kind of activities more accurately since it derives from "wirken," meaning to act, to operate, to influence, to work, to do, to have an effect.

To illustrate it we will shortly describe the development of the child during the *sensorimotor* period. This development can be characterized by different kinds of discoveries the babies make during the first 18 months of life:

They move on a support and discover that there is a world around them - the support, walls, people, which *restrict* their movements. When this happens the moving patterns of the babies change. They elicit changes in resistance between the body and the world - causes and effects - whenever they touch the world around them.

This kind of experience allows them to judge the existence of the world around them and at the same time the existence of their body. With increasing experiences with such causes and effects the babies begin to differentiate what is part of their own body and what is part of the world around them in these cause-effect events.

They discover that there is a *stable support* which does not move, what moves are the parts of their bodies.

Then they discover that there are *"objects" in contrast to the "support"*. When exploring changes of resistance between the body and the support, they may touch "something" on the support. Perhaps that "something" yields their touching movement. In this case, the babies experience a new sequence of changes in resistance, in contrast to, the changes in resistance between the body and support. This experience allows them to consider that "something moves relative to the support which does not move;" the object is perceived in contrast to the support - corresponding to the principle of relativity of perception (Piaget, 1961).

The object becomes important for a child because it allows him to explore *neighboring relationships between object - support - body*, and later on among objects - support - body.

– When babies touch an object on the support,. their fingers move around the ob-

ject, searching for the changes in resistance along its surface, until finally, they embrace the object with all their fingers. When embracing the object, they may be successful in separating the object from the support. Babies experience that an *object can exist separately from the support.* They also learn that the object can be put back on the support and released.
- Babies continue their explorations of objects versus the support and their body by eliciting sequences of changes in resistance, sequences of causes and effects. They touch, separate, and release.
- They begin to differentiate between direct neighboring relationships where object – support – body are touching each other, and indirect ones where one can reach a distant object by crawling across the support.
- They discover the possibility of displacing objects on the support, and thus changing indirect neighboring relationships to direct ones. This means that they start to displace their own body on the support in order to reach a distant object. Such experiences of changing spatial neighboring relationships between body – support – objects are fundamental for the acquisition of spatial concepts.
- By the the middle of the second year children have acquired a critical amount of exploring neigboring relationships so that they can represent them. They start to apply such changes to solve problems of their daily life. This marks the end of the sensorimotor stage

In this development, the object by itself is not important, it becomes important because it serves for "acting upon," for exploring causes and effects.
This is underlined by the concept of "Wirklichkeit".

References

Affolter F (1954) Opérations infralogiques: Comparaisons entre les enfants normaux et sourds. Diplomarbeit, Universität Genf (Unpublished)

Affolter F (1968) Thinking and language. In: International research seminar on the vocational rehabilitation of deaf persons. Dept. Health, Education, and Welfare, Washington D. C.

Affolter F (1970) Developmental aspects of auditory and visual perception: An experimental investigation of central mechanisms of auditory and visual processing. Thesis, Pennsylvania State University (unpublished)

Affolter F (1972) Wahrnehmungsstörungen. In: Schweiz. Taubstummenlehrerverein (Hrsg) Das Mehrfachbehinderte Hörgeschädigte Kind. Marhold, Berlin, S. 94-115.

Affolter F (1974a) Leistungsprofile wahrnehmungsgestörter Kinder. Pädiatr Fortbildung/Klinische Praxis 40: 169-185

Affolter F (1974b) Einsatz und Beschränkung der audiovisuellen Methode im Sprachaufbau des schwer hörbehinderten Kindes. In: Verein Österreichischer Taubstummenlehrer (Hrsg). Audio-visuelle Mittel und Medien im Unterricht Hörgeschädigter. Trauner, Linz

Affolter F (1976) Wahrnehmungsstörungen im Kindesalter. Bull Schweiz. Akad Med Wiss 32: 129-140

Affolter F (1977) Wahrnehmungsgestörte Kinder: Aspekte der Erfassung und Therapie. Pädiatrie Pädol, 12; 205-213

Affolter F, (1981) Jean Piaget: Der Lehrer – der Mensch. Geistige Behinderung (4): 225-229

Affolter F (1985) The development of perceptual processes and problem-solving activities in normal, hearing impaired and language-disturbed children. In: Martin DS (ed) Cognition, education, and deafness. Directions for research and instruction. Gallaudet College Press, Washington D. C.

Affolter F, Bischofberger W (1982) Psychologische Aspekte der Gehörlosigkeit. In: Jussen H, Kröhnert O (Hrsg) Handbuch der Sonderpädagogik, Bd 3. Marhold, Köln, S 605-630

Affolter F, Stricker E (eds) (1980) Perceptual processes as prerequisites for complex human behavior: a theoretical model and its application to therapy. Huber, Bern

Affolter F, Brubaker R, Bischofberger W (1974) Comparative studies between normal and language disturbed children. Acta Otolaryngologica [Suppl] 323

Aslin RN (1981) Development of smooth pursuit in human infants. In: Fisher DF, Monty RA, Senders JW (eds) Eye movements: cognition and visual perception. Erlbaum, Hillsdale, NJ

Ayres AJ (1973) Sensory integration and learning disorders. Western Psychological Services, Los Angeles

Bischofberger W (1989) Aspekte der Entwicklung Taktil-Kinaesthetischer Wahrnehmung: Eine Vergleichsuntersuchung zwischen einer Gruppe sehender und blinder Kinder im Alter von 10 bis 16 Jahren im Taktilen Formerkennen und im Vibrotaktilen Sukzessiven Mustererkennen. Neckar-Verlag, Villingen-Schwenningen

Bloom L, Lahey M (1978) Language development and language disorders. Wiley, New York

Broadbent DE (1958) Perception and communication. Pergamon, Oxford

Broadbent DE (1971) Decision and stress. Academic Press, London

Brown R (1973) A first language: the early stages. George Allen & Unwin, London

Cherry, C (1957) On human communication. MIT Press, Cambridge

Chomsky N (1957) Syntactic structures. Mouton, The Hague

Chomsky N (1965) Aspects of a theory of grammar. MIT Press, Cambridge

Edwards D (1974) Sensory-motor intelligence and semantic relations in early child grammar. Cognition 2: 395-434

Fillmore C (1968) The case for case. In: Bach E, Harms R (eds) Universals in linguistic theory. Holt, Rinehart & Winston, New York

Flavell JH, Markman ED (eds) (1983) Cognitive development. Wiley, New York (Mussen PH (ed) Handbook of child psychology, vol 3)

Forman GE (1982) A search for the origins of equivalence concepts. In: Forman GE (ed) Action and thought: from sensorimotor schemes to symbolic operations. Academic Press, New York

Fraiberg S (1977) Insights from the blind: comparative studies of blind and sighted infants. Basic Books, New York

Furth H (1966) Thinking without language: psychological implications of deafness. Free press, New York

Gardner H (1983) Frames of mind. Basic Books, New York

Howard IP, Templeton WB (1966) Human spatial orientation. Wiley, New York

Jakobson, R (1969) Kindersprache, Aphasie und allgemeine Lautgesetze. Edition Suhrkamp, Frankfurt/M.

Katz D (1948) Gestaltpsychologie. Schwabe, Basel

Koffka K (1963) Principles of gestalt psychology. Harcourt, Brace & World, New York

Lashley K (1951) The problem of serial order in behavior. In: Jeffries LA (ed) Cerebral mechanisms in behavior. Wiley, New York

Menyuk P (1971) The acquisition and development of language. Prentice Hall, Englewood Cliffs, NJ

Miller GA (1956) The magical number seven, plus or minus two. Some limits on our capacity for processing information. Psychological Review 63: 81-97

Miller GA (1967) The psychology of communication. Penguin Books, Baltimore

Neisser U (1976) Cognition and reality. Freeman, San Francisco

Norman DA (1982) Learning and memory. Freeman, San Francisco

Palermo DS (1978) Psychology of language. Scott Foresman, Glenview, Ill

Palermo DS, Molfese DL (1972) Language acquisition from age five onward. Psychological Bulletin 78: 409-428

Piaget J (1950) Psychology of intelligence. Harcourt Brace, New York, Original: La Psychologie de l'Intelligence. Librairie Armand Collin, Paris 1947

Piaget J (1952) The origins of intelligence in children. International Universities Press, New York. Original: La Naissance de l'Intelligence chez l'Enfant. Delachaux et Niestlé, Neuchâtel 1936

Piaget J (1958) Introduction. In: Jonckheere A, Mandelbrot B, Piaget J (eds) La lecture de l'expérience. Bibl. Scientif. Internat., Etudes d'épistémologie génétique. PUF, Paris pp 1-9

Piaget J (1961) Les mécanismes perceptifs: Modèles probabilistes, analyse génétique, relations avec l'intelligence. PUF, Paris

Piaget J (1962) Play, dreams and imitation in childhood. Norton, New York. Original: La Formation du Symbole chez l'Enfant. Delachaux et Niestlé, Neuchâtel 1945

Piaget J (1963) Le language et les opérations intellectuelles. In: Problèmes de psycholinguistique: Symposium de l'association de psychologie scientifique de langue française. PUF, Paris

Piaget J, Inhelder B (1956) The child's conception of space. Routledge & Paul, London. Original: La Représentation de L'Espace. PUF, Paris 1948

Piaget J, Morf A (1958) Les préinférences perceptives et leurs relations avec les schèmes sensori-moteurs et opératoires. In Bruner J, Bresson F, Morf A, Piaget J (eds) Logique et perception. Bibl. Scientif. Internat., Etudes d'épistémologie génétique. PUF, Paris, pp 117-156

Pick HL (1980) Tactual and haptic perception. In: Blasch BB, Welsh RL (eds) Orientation and mobility for visually handicapped persons: development and fundamental principles. American Foundation for the Blind, New York

Premack D (1970) The education of Sarah: a chimp learns language. Psychology Today 4/4: 55-58

Premack D (1971) Language in the chimpanzee? Science 172: 808-822

Premack D (1975) Intelligence in ape and man. Holt, Rinehart & Winston, New York

Riesen AH (1975) The developmental neuropsychology of sensory deprivation. Academic Press, New York

Salapatek P, Cohen LB (eds) (1986) Handbook of infant perception. Vol. 1: From sensation to perception. Vol. 2: From perception to cognition. Academic Press, New York

Schiff W, Foulke E (eds) (1982) Tactual perception: a sourcebook. University Press, Cambridge

Stambak M (1963) Tonus et psychomotricité dans la petite enfance. Delachaux et Niestlé, Neuchâtel

Terrace HS (1979) Nim, a chimpanzee who learned sign language. Washington Square Press, New York

Subject Index

act upon (*see also* cause-effect, information, mouth, perceive) 38, 189–191, 193–197, 211, 237, 257, 292
aggressive *see* behavior
anticipation (*see also* interiorize, wait) 26, 28, 187, 222, 245, 252, 267
–, definition 177
aphasia (*see also* language) 285, 286, 290
articulation *see* speech
attention (*see also* capacity) 251
audio-visual (*see also* intermodal) 292
auditory (*see also* hearing, information, perceive) 153, 160, 163
– effects 139, 140
autistic 86, 293

babble *see* speech
behavior (*see also* hyperactive)
–, aggressive 92–94, 152, 154, 293
–, ill-mannered 94, 95
–, self-destructive 94
–, talkative 89–92
bimanual (*see also* hands) 295
bite *see* mouth
blind *see* visually impaired
body (*see also* event, support, surrounding, touch) 8–10, 14, 179, 196, 210, 215, 255, 270, 271, 281
– position 188, 189, 196
–, taking off 44
–, tense 100
– tone (*see also* spastic)
– –, changes 170–172, 177, 290
– –, hypertonic 171
– –, hypotonic 125, 126, 141, 171, 184
brain damage 86, 91–93, 108, 113, 127, 133, 142, 172, 247, 292, 293
– lesion *see* brain damage

capacity 151–153
– bundle 153
–, overload 154, 293
case grammar *see* grammar
cause-effect (*see also* information, problem solving, resistance, tool) 3, 38, 48, 135–137, 147, 163, 169, 175, 193, 203, 204, 206, 208, 219, 220, 237, 241, 247, 255, 260, 262, 280, 282, 292
cerebral palsy 170, 172, 182, 212
change in resistance *see* resistance
cognitive (*see also* cause-effect, concept formation, interaction, root) 8, 179

communication 290
compensation 153
competence (*see also* performance, production, situation) 227, 250
comprehension *see* understanding, language comprehension
concept formation 48, 255
contact *see* social
coordination *see* intermodal
crawling 6
curiosity 250, 251, 260

daily *see* event
deaf *see* hearing impaired
decision 248
deep structure 279, 280, 283, 290
detour (*see also* planning, representation) 250, 252, 256, 257, 263
development (*see also* hands, language, perceive, problem solving, release, root, tactile-kinesthetic)
–, deviant 203, 281–283
–, level 160, 162, 164, 165
–, origin 162, 163, 292
–, prerequisite 160
deviant *see* development, fingers, information, world
displace *see* move
distance (*see also* neighboring relationships) 131, 149, 150, 203
dominance (*see also* hands) 33, 181
drawing *see* symbolic
dyslexic 86

eat *see* habit
embrace (*see also* fingers, hold, mouth, object, take) 8, 9, 28, 31, 192
–, mouth 17, 18, 29
–, niche 8–12, 14, 113
– object 16, 120, 122
emotional (*see also* interaction, root) 8, 80–82, 179
– disorder 86
environment (*see also* intervention, surroundings) 292
event (*see also* guide, interiorize, problem solving) 227, 273, 275, 277, 278, 286, 290
–, daily 70–72, 220, 227, 234, 242, 256, 257, 269–271, 279, 280, 284, 285
–, new 29, 30, 251
explore *see* mouth

303

family life *see* event, daily
fear   134, 147
feel *see* tactile-kinesthetic, touch
figure-ground   41
fingers (*see also* embrace, resistance, support, take, touch)
-, deviant   149
-, hold   118-121, 123, 124
-, touch   97-99, 123, 149, 194
form *see* semiotic
frustration   153, 154

gaze *see* visual
geriatric   292, 295
goal   77, 163, 209, 263
-, intervention   245
grammar   277, 279
-, case grammar   280
-, kernel sentence   280, 283
-, syntactic rules   283
-, transformation rules   280
grasp *see* embrace, fingers, take
gravity *see* support
guide (*see also* act upon, neighboring relationships, problem solving, side, support, tactile-kinesthetic, understanding)   177, 193, 235, 236, 242, 253
-, difficulty   234
- hands   192, 198, 201, 253, 255
-, how   178-184, 197
-, spontaneous   166, 167
-, talk   236, 286

habit   69, 236, 248-252, 257, 292
-, eat   184
-, orderly behavior   70
hands (*see also* bimanual, embrace, guide, support, take)   180, 196, 216
-, development   31-33, 148, 197, 210
-, disturbances   115-117
head control   172
- trauma (*see also* brain damage)   86, 93, 133, 172, 173, 251, 284
hearing (*see also* auditory)   23, 242, 276, 277, 281
-, impaired   XV, 160-164, 268, 276
hide *see* world
hold *see* embrace, resistance, take
hyperactive (*see also* behavior)   150, 220, 252
-, hectic   88, 246, 247
hypertonic *see* body tone
hypotonic *see* body tone

imitation (*see also* semiotic, sensorimotor, symbolic)   164, 276, 283
-, deferred   160
-, direct   161
information (*see also* auditory, capacity, object, release, situation, tactile-kinesthetic, visual)   211, 250-252
-, deviant   144, 145
-, kind   139, 141, 142, 199

-, search   129, 146, 151, 153
inhibition   22, 29
interaction (*see also* cognitive, emotional, tactile-kinesthetic)   22
-, definition   5
- experience   145, 147
-, tactile-kinesthetic   163-165, 179, 189, 276, 292
interiorize (*see also* language, representation, semiotic, symbolic)   164, 244, 247, 290
- events   245, 256, 267, 268
intermodal (*see also* audio-visual, modality-specific, perceive)
- coordination   25, 28, 30, 52, 100, 140, 152, 163, 171, 173, 174, 175
- disorder   161
intervention (*see also* goal, learning, motivation, problem solving, root)   138, 164, 165, 253, 284, 293
-, environment   169
-, occupational therapy   173
-, physical therapy   173
-, progress   193
-, skill   164, 165
-, therapy   241, 242, 247

jerk *see* tactile-defensive, withdraw

kernel sentence *see* grammar
knock   125

language (*see also* development, interiorize, representation, semiotic, symbolic)   160, 280
- comprehension (*see also* understanding)   161, 268, 275, 283, 286, 290
-, creativity   279, 280, 283
-, definition   164, 268, 276
- development   160, 161, 285
- discovery   160, 161, 164, 268
- disorder *see* aphasia
- forms *see* semiotic
-, production   283, 286
- therapy *see* aphasia, intervention
- universals   279
learning (*see also* habit, interaction, intervention, operative, problem solving, production, root, tactile-kinesthetic, understanding)   250, 251, 290, 292
-, definition   167, 169
-, disordered   86
-, levels   170, 171, 178, 203, 222, 252
-, prerequisites   227
-, regression   247, 252, 253
-, tactile-kinesthetic   166
level *see* development, learning
look (*see also* visual)   22-25

memory *see* retrieve
modality-specific (*see also* auditory, intermodal, perceive, visual)   22, 30, 174
motion   247, 248
motivation   253

motor 148, 172, 181
mouth (*see also* embrace, guide, object, perceptually disturbed) 16-19, 27, 29, 126, 182, 194
-, act upon 184
-, bite 126, 184, 186
-, explore 184, 187
move (*see also* resistance) 247, 253
-, displace 192
- support 40, 41
- touch 38, 253

neighboring relationships (*see also* guide, support, tactile-kinesthetic) 122, 148, 163, 253
-, conditions 130, 150, 151
-, direct/indirect 197, 203
-, production 128, 129
-, take out/put in 46, 47
-, through 49-51
niche *see* embrace, resistance, side, support

object (*see also* embrace, information, neighboring relationships, recognize, tactile-kinesthetic, take, touch)
-, mouthing 17, 18
-, qualities 189, 211
-, support 16, 46, 120, 148, 149, 203
operative 250

perceive (*see also* auditory, modality-specific, intermodal, motor, sequences, tactile-kinesthetic, visual)
-, act upon 73-77, 122
- cause-effect 38
-, definition 3, 28
-, development 163, 164
- relativity 196
perceptually disordered (*see also* development, information, language, root, symbolic) 86, 161-165, 251, 252, 281-285, 293,
- mouthing (*see also* object) 182, 184
performance *see* competence, production
picture *see* symbolic
position *see* body
pretend *see* symbolic
problem solving (*see also* cause-effect, event, resistance, root, sequence) 147, 165, 195, 196, 219, 220, 228, 230, 233, 235-237, 241, 253, 271
- daily living 72, 77, 163-165, 174, 189, 192, 203, 228, 257, 262, 285, 290, 292, 293
-, description 77, 162
-, development 162, 163
- disorder 162
problems 227
product 234, 235
production (*see also* habit, language, understanding) 245-247, 251, 256, 257, 267
-, level 222, 227, 251-253, 290
progress *see* learning
psychiatric 293, 295
-, changes in personality 253

read *see* semiotic
recognize (*see also* anticipation) 175, 222, 245
- action 275, 282
-, definition 177
- object 275, 282
-, visual 101, 252, 275
reference *see* support
relativity *see* perceive
release (*see also* embrace, information, resistance, support, tactile-kinesthetic, take, touch) 199, 202
-, development 29, 34, 46
-, disturbances 127, 148, 150
representation (*see also* detour, interiorize, language, semiotic, symbolic) 147, 256, 267-271, 273, 275, 281, 283, 285, 286, 290
- definition 267
- planning 247, 252, 263
resistance (*see also* embrace, object, side, support, take off) 3, 191, 196, 229, 291
-, change 5-7, 16, 35, 39, 40, 120, 125, 148, 149, 192-194, 202, 204, 206, 208, 209, 219, 224, 230, 243, 253, 255, 266, 272
-, maximal changes 42, 53, 108, 113, 127, 128, 142, 144, 146, 192, 253
retrieve 146, 147, 273, 284, 290
-, memory disorder 153
root (*see also* development) 163-165, 276, 279, 280, 284, 292, 293
rule (*see also* act upon, take, touch, side, support)
-, taking off 41, 44, 47, 122-125, 163
-, touching 6, 96-104, 163, 196, 272
-, side 6, 8, 163
-, support 6, 105-108, 163

search *see* information
see (*see also* visual) 26-28, 143, 242, 281
semiotic (*see also* interiorize, language, representation, symbolic) 160, 164, 275, 283
-, definition 267, 268, 276, 285
- form 267, 268, 271, 273, 279, 286, 290
- production 285
-, reading 279, 290
- sign 268, 279, 285
-, writing 279, 290
sensorimotor (*see also* imitation, perceive) 160, 163, 164
- signal 163, 275, 283, 285
separate (*see also* release, rule of taking off) 42
- support 41
serial *see* sequences
sequences (*see also* capacity) 131-133, 151, 152, 208, 220, 245, 248 250, 252, 255
-, disorder 161
-, serial 131
side (*see also* resistance, rule of touching, touch) 8, 11, 198
- search 109-111
-, stable 8, 13, 190, 254
sign *see* semiotic
signal *see* sensorimotor

305

situation   138, 139, 142, 148, 152–154, 165, 189, 249–252, 266, 273, 279, 282
–, variety   174
social (*see also* withdraw)   80–82, 147
– contact   113, 114
spastic (*see also* body tone)   86, 171, 293
speech (*see also* language, semiotic, sensorimotor)
–, babble   160, 162, 276, 283
– learning   276, 277, 283, 284
–, prerequisites   19
– sounds   162, 267, 290
stability *see* support
stage *see* development
stress (*see also* capacity)   293
support (*see also* embrace, move, object, release, resistance, rule, separate, take off)   5, 6, 11, 146, 203, 213
–, gravity   5, 6
–, reference   148–150, 196, 198, 199, 243, 253
–, stable   7, 8, 13, 104, 190, 196, 223, 224, 254
surface structure   279, 280, 283, 290
surrounding (*see also* body, embrace, mouth, three-dimensional, touch, world)   3, 8, 31, 137, 150, 163, 189, 191, 193, 255, 257
–, destruction   220
symbol   267, 268, 275, 282
symbolic (*see also* imitation, interiorize, language, representation, semiotic)   268, 273
– drawing   162, 276, 283, 285, 286
– picture   275, 282, 285
– role playing   270
– play   275, 281
syntactic rules *see* grammar

tactile-defensive (*see also* withdraw)   114, 115
tactile-kinesthetic (*see also* body, information, interaction, learning, perceive, touch)   23–25, 145–147, 153, 163, 165–167, 179, 188, 190, 192, 193, 222, 227, 231, 235–237, 241, 253, 255, 276, 279, 286, 290, 293
–, feel   13, 23–25, 145, 242
– development   166
– disorder   161–163, 252, 267
–, physiology   166
take (*see also* embrace, information, mouth, object, release, resistance, rule, tactile-kinesthetic)   28
–, grasp   26, 27

– off   38, 41, 43, 122–125, 128, 192, 247, 253
– out *see* neighboring relationships
talk *see* guide
temporal-successive *see* sequence
therapy *see* intervention
three-dimensional (*see also* embrace, surroundings, world)   21, 120, 150
toilet-training   96, 142
tool (*see also* hands, guide)   202, 210, 216, 255
– knife   217, 219
touch (*see also* act upon, embrace, fingers, move, release, rule, support, surroundings, take, tactile-kinesthetic)   166, 172, 192, 247, 272
–, body   18, 146
–, mouth   19
– object   16
–, variety   33–36
–, world   3, 28
transfer   138, 173, 250
transformation rules *see* grammar

understanding (*see also* language comprehension, production)   167, 227, 245, 257
–, definition   221, 222
–, interest   221
–, level   222, 227, 236, 283

verbal   252, 268, 275, 276, 278, 279, 283
visual (*see also* look, see)   145–147, 150, 153, 160, 163, 166, 235, 252, 282, 283
– effect   139, 140
– look away   99, 100, 152
visually impaired   160, 161, 163, 164

wait (*see also* anticipation)   133, 134, 147
walk   101–104, 148, 173
– on stairs   105–108
withdraw (*see also* social, tactile-defensive)   28, 96–98
world (*see also* surroundings, three-dimensional, touch)   3, 4, 6, 10, 22, 150
–, deviant   150
–, hide   14
–, unity   33
wrap   53, 54
writing *see* semiotic

P. M. Davies, Bad Ragaz, Switzerland

Right in the Middle

Selective Trunk Activity in the Treatment of Adult Hemiplegia

Foreword by S. Klein-Vogelbach

1990. XVIII, 274 pp. 316 figs. Softcover DM 64,–
ISBN 3-540-51242-X

This book is a must for everyone involved in the rehabilitation of stroke patients. Certainly all those already familiar with Pat Davies' best-selling guide to the treatment of adult hemiplegia **Steps to Follow** will want to learn more about the treatment advances presented here. The new book focuses on a subject that has been almost completely ignored until now in the rehabilitation of hemiplegic patients: selective trunk activity.

The author once again shares her vast experience in treating patients with hemiplegia and points to the impressive results that can be achieved when specific therapy to retrain and regain selective trunk activity is integrated into the treatment program. The key to successful treatment lies in regaining adaptive stabilisation of the trunk, and the ability to move parts of it in isolation.

The book explains how the loss of trunk control causes difficulties with breathing, speaking, balance and walking, as well as functional use of the arm and hand. Activities to improve these abilities are described.

Clear concise instructions, illustrated by a wealth of photographs of patients in action, will help doctors, physiotherapists and occupational therapists to observe, analyse and overcome the problems caused by inadequate trunk control.

Springer-Verlag
Berlin Heidelberg
New York London Paris
Tokyo Hong Kong

S. Klein-Vogelbach, Bottmingen, Switzerland

Functional Kinetics

Observing, Analyzing, and Teaching Human Movement

Translated from the German by G. Whitehouse

1990. XIV, 337 pp. 329 figs. 1 tab.
Softcover DM 68,– ISBN 3-540-15350-0

Susanne Klein-Vogelbach's acclaimed textbook describing her concept of functional kinetics is now available in English! The well-known physiotherapist has once again revised the material to improve the presentation of her theory concerning the systematic observation and analysis of human movement. The direct observation of movement is an excellent foundation on which to build successful therapeutic programs. The ideas outlined in this book are basic to physical therapy and rehabilitation and should be familiar to every active therapist.

Springer-Verlag
Berlin Heidelberg
New York London Paris
Tokyo Hong Kong

Prices are subject to change without notice.